René Erlín Castillo, Héctor Camilo Chaparro Gutiérrez, Julio César Ramos-
Functional and Harmonic Analysis

Also of Interest

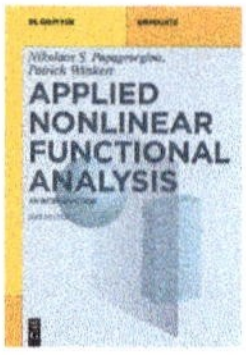

Applied Nonlinear Functional Analysis
An Introduction
Nikolaos S. Papageorgiou, Patrick Winkert, 2024
ISBN 978-3-11-128421-7, e-ISBN (PDF) 978-3-11-128695-2

Harmonic Analysis and Convexity
Edited by Alexander Koldobsky, Alexander Volberg, 2023
ISBN 978-3-11-077537-2, e-ISBN (PDF) 978-3-11-077538-9
in
Advances in Analysis and Geometry
ISSN 2511-0438

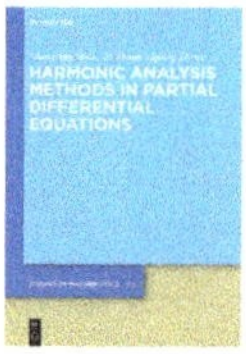

Harmonic Analysis Methods in Partial Differential Equations
Changxing Miao, Bo Zhang, Jiqiang Zheng, 2025
ISBN 978-3-11-138451-1, e-ISBN (PDF) 978-3-11-138472-6
in
De Gruyter Studies in Mathematics
ISSN 0179-0986

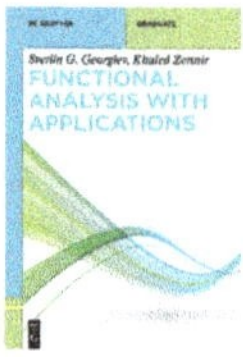

Functional Analysis with Applications
Svetlin G. Georgiev, Khaled Zennir, 2019
ISBN 978-3-11-065769-2, e-ISBN (PDF) 978-3-11-065772-2

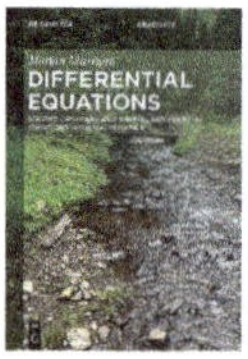

Differential Equations
Solving Ordinary and Partial Differential Equations with Mathematica®
Marian Mureşan, 2024
ISBN 978-3-11-141109-5, e-ISBN (PDF) 978-3-11-141139-2

René Erlín Castillo, Héctor Camilo Chaparro Gutiérrez, Julio César Ramos-Fernández

Functional and Harmonic Analysis

Weighted Lebesgue Spaces, Muckenhoupt Weights, Convexity and BMO Spaces

DE GRUYTER

Mathematics Subject Classification 2020
Primary: 46E30, 26A51, 42B20; Secondary: 42B25, 28A12, 30H35, 42B35

Authors
Prof. Dr. René Erlín Castillo
Universidad Nacional de Colombia
Departamento de Matemáticas
Bogotá 111321
Colombia
recastillo@unal.edu.co

Prof. Dr. Héctor Camilo Chaparro Gutiérrez
Universidad de Cartagena
Programa de Matemáticas
Cartagena de Indias 130014
Colombia
hchaparrog@unicartagena.edu.co

Prof. Dr. Julio César Ramos-Fernández
Universidad Distrital Francisco José de Caldas
Facultad de Ciencias Matemáticas y Naturales
Bogotá 111711
Colombia
jcramosf@udistrital.edu.co

ISBN 978-3-11-914220-5
e-ISBN (PDF) 978-3-11-222324-6
e-ISBN (EPUB) 978-3-11-222364-2

Library of Congress Control Number: 2025945214

Bibliographic information published by the Deutsche Nationalbibliothek
The Deutsche Nationalbibliothek lists this publication in the Deutsche Nationalbibliografie; detailed bibliographic data are available on the Internet at http://dnb.dnb.de.

Cover image: fotografstockholm / E+ / Getty Images
Typesetting: VTeX UAB, Lithuania

www.degruyter.com
Questions about General Product Safety Regulation:
productsafety@degruyterbrill.com

A la memoria de Julieta mi mamá-abuela
A mi esposa Hilcia del Carmen
A mis hijos: René José
Manuel Alejandro
Irene Gabriela
Renzo Rafael

R.E.C.

A mis padres: Blanca Cecilia
Héctor
A mis hermanos: Diego Fernando
Oscar Alejandro
A mi hijo: David Camilo

H.C.Ch.

A la memoria de mis padres: Onelia
César
A mi esposa: Margot
A mi hijo: Carlos Julio

J.C.R-F.

Contents

Introduction

In the vast landscape of functional analysis, a realm where mathematical structures interplay with real-world phenomena, the study of function spaces holds a central position. Among these spaces, the Lorentz spaces emerge as a cornerstone, providing a nuanced understanding of the behavior of functions concerning their growth rates and distribution. Weighted Lorentz spaces, a refined extension of the classical Lorentz spaces, constitute a fascinating domain where the interplay between weights and integrability properties enriches the analytical landscape. These spaces, named after the mathematician G. G. Lorentz, are a natural extension of the classical Lebesgue spaces, allowing for a more nuanced understanding of the behavior of functions in different settings. In this book, we will delve into the world of weighted Lorentz spaces, exploring their properties, applications, and connections to other areas of mathematics.

The genesis of weighted Lorentz spaces can be traced back to the pioneering works of Hardy and Littlewood, who laid the foundation for the theory of function spaces with integral norms. The classical Lorentz spaces, introduced by Lorentz in the mid-twentieth century, were initially formulated as a means to generalize the L_p spaces, allowing for a more flexible treatment of functions with varying rates of growth. These spaces garnered significant attention due to their capacity to capture the intricate behavior of functions near the boundary of their domains.

In the 1950s and 1960s, mathematicians like A. Zygmund and E. Stein further expanded the theory of weighted Lorentz spaces, introducing new techniques and results that have had a lasting impact on the field. The 1970s and 1980s saw a surge in research on weighted Lorentz spaces, with mathematicians like C. Fefferman, R. Fefferman, and S. Wainger making significant contributions to the field.

Building upon this legacy, the concept of weighted Lorentz spaces emerged as a natural generalization, incorporating the influence of weights into the norm structure. The introduction of weights enables a more nuanced characterization of functions, reflecting the underlying distributional properties and emphasizing certain regions of the domain over others. This augmentation facilitates a deeper understanding of functions in contexts where nonuniform growth rates or localized behavior play a crucial role.

The weighted norm inequalities have became one of the most dynamically developing parts of harmonic analysis since the early 1970s and the pioneering result by B. Muckenhoupt [58].

Solutions of many important problems have been closely linked with weight problems. The mentioned paper by B. Muckenhoupt triggered a flood of results on weighted inequalities and related topics; in that paper it was shown among other things that the one weighted norm inequality for the maximal operator is true if and only if the weight satisfies the so called A_p condition.

https://doi.org/10.1515/9783112223246-204

The A_p weights provide an extraordinarily beautiful answer to a number of challenging problems which have arisen already in the 1930s in connection with the fundamental results due to G. H. Hardy and J. E. Littlewood [43, 45]. Theorems on boundedness of the Hilbert transform, fractional-order maximal functions, fractional integrals followed very soon; see R. A. Hunt, B. Muckenhoupt, and R. L. Wheeden [48], B. Muckenhoupt and R. L. Wheeden [59], R. R. Coifman and C. Fefferman [27].

It is perhaps impossible to give a full account of related problems and to trace back the (by no means straightforward) way to the first celebrated papers on weighted inequalities in the early 1970s. Several branches of analysis meet here, influencing each other—one could mention a very large number of papers—and still the list would be far from complete.

It is important to note that in the year 1985 the monograph of J. García-Cuerva and J. Rubio de Francia [37] and a year later that of A. Torchinsky [66] appeared, which are excellent sources on the subject; also, a large bibliography can be found in them.

A particularly interesting topic covered in the latter book is the study of Bounded Mean Oscillation (BMO) spaces, which are discussed in Chapter VIII. The BMO spaces play a fundamental role in modern analysis, particularly in the study of singular integral operators and PDEs. These spaces provide a natural setting for understanding functions whose oscillations are locally controlled, making them a crucial tool in harmonic analysis. The chapter explores the classical John–Nirenberg inequality, which establishes the exponential integrability of BMO functions, as well as the completeness properties of these spaces. Furthermore, BMO spaces serve as a bridge between L^p and Hardy spaces, offering a powerful framework for interpolation and applications in various branches of analysis.

Our first intention in writing these notes was to create a seminar to prepare a few students to begin doing research on weighted Lebesgue spaces. The original manuscript was largely written in the style of an oral class. Although this presentation is more or less formal, the emphasis is still on the didactic aspect because now we target those students who are not generally able to attend regular courses in mathematics centers. Over time the initial objective of this project was reflected in four research papers (see [12, 13, 15, 20]).

Also, under the supervision of the first author, the second author continued doing research on this topic until completing his PhD dissertation (see [11, 23]).

Part of this book content has been tested with students from Universidad Nacional de Colombia, Universidad Distrital Francisco José de Caldas, as well as Universidad de Cartagena in the classes on advanced topics in analysis and also on measure theory and integration.

This book is intended for graduate students and researchers interested in weighted function spaces and their applications. The reader is expected to have a solid foundation in real and functional analysis, as these areas form the backbone of the discussions presented here.

The authors extend their sincere gratitude to the entire team at De Gruyter for their role in publishing this book. We are particularly indebted to Editor Ranis Ibragimov, whose dedication and support were indispensable. We also extend our thanks to Production Manager Vilma Vaičeliūnienė for her invaluable assistance during the publication process.

Bogotá, Colombia, September 2025	René Erlín Castillo
Cartagena de Indias, Colombia, September 2025	Héctor Camilo Chaparro Gutiérrez
Bogotá, Colombia, September 2025	Julio César Ramos-Fernández

1 The $\ell_{p,\omega}$ space

The study of weighted Lebesgue sequence spaces provides fundamental insights into the behavior of sequences under the influence of weights, extending the classical theory of Lebesgue sequence spaces. These spaces serve as a crucial bridge to understanding more advanced function spaces, such as weighted Lorentz spaces, where weights introduce refined integrability and convergence properties. By examining the structural and functional properties of weighted sequence spaces, we gain deeper intuition about how weights modulate the analytical and geometric features of these spaces.

In this chapter, we introduce the weighted Lebesgue sequence space using a weight sequence. We explore key properties of these spaces, including Hölder and Minkowski inequalities, completeness, separability, and embedding results. Additionally, we study their strict and uniform convexity, as well as the duality theory of these spaces.

We remark that the content of Chapters 1 and 2 originates from the master's thesis [24], with author René Castillo serving as the thesis advisor.

1.1 The weighted Lebesgue sequence space

Definition 1.1 (Lebesgue sequence space). The *Lebesgue sequence space* (also known as the *discrete Lebesgue space*) with $1 \leq p < \infty$, denoted by ℓ_p or sometimes also by $\ell_p(\mathbb{N})$, stands for the set of all sequences of real numbers $\mathbf{x} = (x_n)_{n\in\mathbb{N}}$ such that $\sum_{k=1}^{\infty} |x_k|^p < \infty$, i. e.,

$$\ell_p := \left\{ \mathbf{x} = (x_n)_{n\in\mathbb{N}} : \sum_{n=1}^{\infty} |x_n|^p < \infty,\ x_n \in \mathbb{R},\ \forall n \in \mathbb{N} \right\}.$$

For a sequence $\mathbf{x} = (x_n)_{n\in\mathbb{N}} \in \ell_p$, its norm is given by

$$\|\mathbf{x}\|_p := \left(\sum_{n=1}^{\infty} |x_n|^p \right)^{\frac{1}{p}}.$$

For $p = \infty$, the set ℓ_∞ is the set of bounded real sequences and its norm is given by

$$\|\mathbf{x}\|_\infty := \sup\{|x_n| : n \in \mathbb{N}\}.$$

Definition 1.2. A weight sequence $\boldsymbol{\omega} = (\omega(n)) = (\omega_n)$ is a positive sequence.

By joining the previous concepts, we can now define the *weighted Lebesgue sequence space*.

Definition 1.3. Let $1 \leq p < +\infty$. For a fixed p and a fixed weight sequence $\omega = \omega_n$, the set $\ell_{p,\omega}$ is defined as follows:

$$\ell_{p,\omega} := \left\{ \mathbf{x} = (x_n)_{n\in\mathbb{N}} : \sum_{n=1}^{\infty} |x_n|^p \omega_n < \infty,\ x_n \in \mathbb{R},\ \forall n \in \mathbb{N} \right\},$$

https://doi.org/10.1515/9783112223246-001

with

$$\|\mathbf{x}\|_{p,\omega} := \left(\sum_{n=1}^{\infty} |x_n|^p \omega_n\right)^{\frac{1}{p}}.$$

For $p = \infty$, we define $\ell_{\infty,\omega}$ to be the set of all bounded real sequences, with the norm

$$\|\mathbf{x}\|_{\infty,\omega} := \sup\{|x_n| : n \in \mathbb{N}\}.$$

Thus $\ell_{\infty,\omega}$ coincides with the familiar unweighted space ℓ_∞. Also, note that the weighted Lebesgue sequence space is defined by considering the weight as a multiplier, that is,

$$\mathbf{x} \in \ell_{p,\omega} \iff \mathbf{x}\omega^{1/p} \in \ell_p. \tag{1.1}$$

It is easy to show that the set $\ell_{p,\omega}$ is a *real vector space*. Indeed, note that for $a, b \in \mathbb{R}$, we have

$$\begin{aligned} |a+b|^p &\leq (|a| + |b|)^p \\ &\leq (2\max\{|a|, |b|\})^p \\ &= 2^p \max\{|a|^p, |b|^p\} \\ &\leq 2^p(|a|^p + |b|^p). \end{aligned}$$

Let $\lambda \in \mathbb{R}$ and $\mathbf{x}, \mathbf{y} \in \ell_{p,\omega}$, then

$$\begin{aligned} \sum_{n=1}^{\infty} |\lambda x_n + y_n|^p \omega_n &= \sum_{n=1}^{\infty} |\lambda x_n \omega_n^{1/p} + y_n \omega_n^{1/p}|^p \\ &\leq \sum_{n=1}^{\infty} 2^p(|\lambda x_n \omega_n^{1/p}|^p + |y_n \omega_n^{1/p}|^p) \\ &= 2^p\left(\sum_{n=1}^{\infty} |\lambda|^p |x_n|^p \omega_n + |y_n|^p \omega_n\right) \\ &= 2^p\left(|\lambda|^p \sum_{n=1}^{\infty} |x_n|^p \omega_n + \sum_{n=1}^{\infty} |y_n|^p \omega_n\right) \\ &< \infty. \end{aligned}$$

Thus $\lambda\mathbf{x} + \mathbf{y} \in \ell_{p,\omega}$. The remaining properties for $\ell_{p,\omega}$ to be a vector space are trivial.

Now, we are going to study basic properties and inequalities of the $\ell_{p,\omega}$ space by using the theory of the ℓ_p spaces. We begin by deriving the classic Hölder's inequality.

Theorem 1.1 (Weighted Hölder's inequality). *Let $1 < p, q < +\infty$ with $\frac{1}{p} + \frac{1}{q} = 1$ and $\mathbf{x} = (x_n)_{n\in\mathbb{N}} \in \ell_{p,\omega}$, $\mathbf{y} = (y_n)_{n\in\mathbb{N}} \in \ell_{q,\omega}$. Then*

$$\sum_{k=1}^{\infty} |x_k y_k| \omega_k \leq \left(\sum_{k=1}^{\infty} |x_k|^p \omega_k\right)^{1/p} \left(\sum_{k=1}^{\infty} |y_k|^q \omega_k\right)^{1/q},$$

i. e.,

$$\sum_{k=1}^{\infty} |x_k y_k| \omega_k \leq \|x\|_{p,\omega} \|y\|_{q,\omega}.$$

Proof. Let $\mathbf{x} \in \ell_{p,\omega}$, $\mathbf{y} \in \ell_{q,\omega}$. Then

$$\begin{aligned}
\sum_{k=1}^{\infty} |x_k y_k| \omega_k &= \sum_{k=1}^{\infty} |x_k y_k| \omega_k^{\frac{1}{p}+\frac{1}{q}} \\
&= \sum_{k=1}^{\infty} |x_k| \omega_k^{\frac{1}{p}} |y_k| \omega_k^{\frac{1}{q}} \\
&= \sum_{k=1}^{\infty} \underbrace{|x_k \omega_k^{\frac{1}{p}}|}_{\in \mathbb{R}} \underbrace{|y_k \omega_k^{\frac{1}{q}}|}_{\in \mathbb{R}} \\
&\leq \left(\sum_{k=1}^{\infty} |x_k \omega_k^{\frac{1}{p}}|^p \right)^{1/p} \left(\sum_{k=1}^{\infty} |y_k \omega_k^{\frac{1}{q}}|^q \right)^{1/q} \qquad \text{(by Hölder inequality in } \ell_p) \\
&= \left(\sum_{k=1}^{\infty} |x_k|^p \omega_k \right)^{1/p} \left(\sum_{k=1}^{\infty} |y_k|^q \omega_k \right)^{1/q} \\
&= \|\mathbf{x}\|_{p,\omega} \|\mathbf{y}\|_{q,\omega}.
\end{aligned}$$

□

Now, we will derive the Minkowski inequality, by using the Hölder's inequality. It is important to note that our approach is not the only way possible; see [10, 64].

Theorem 1.2 (Weighted Minkowski inequality). *Let* $\mathbf{x} = (x_n)_{n\in\mathbb{N}}, \mathbf{y} = (y_n)_{n\in\mathbb{N}} \in \ell_{p,\omega}$ *and* $p \geq 1$. *Then*

$$\left(\sum_{k=1}^{\infty} |x_k + y_k|^p \omega_k \right)^{1/p} \leq \left(\sum_{k=1}^{\infty} |x_k|^p \omega_k \right)^{1/p} + \left(\sum_{k=1}^{\infty} |y_k|^p \omega_k \right)^{1/p},$$

i. e.

$$\|\mathbf{x} + \mathbf{y}\|_{p,\omega} \leq \|\mathbf{x}\|_{p,\omega} + \|\mathbf{y}\|_{p,\omega}.$$

Proof. Suppose that $\mathbf{x} = (x_n)_{n\in\mathbb{N}}, \mathbf{y} = (y_n)_{n\in\mathbb{N}} \in \ell_{p,\omega}$, and $\frac{1}{p} + \frac{1}{q} = 1$. Then, we have

$$\sum_{k=1}^{\infty} |x_k + y_k|^p \omega_k = \sum_{k=1}^{\infty} |x_k + y_k|^{p-1} |x_k + y_k| \omega_k,$$

and for any $k \in \mathbb{N}$,

$$|x_k + y_k|^{p-1} |x_k + y_k| \leq |x_k + y_k|^{p-1} |x_k| + |x_k + y_k|^{p-1} |y_k|.$$

Then,

$$\sum_{k=1}^{\infty}|x_k+y_k|^p\omega_k \le \sum_{k=1}^{\infty}|x_k+y_k|^{p-1}|x_k|\omega_k + \sum_{k=1}^{\infty}|x_k+y_k|^{p-1}|y_k|\omega_k,$$

and, applying Theorem 1.1 properly, we have

$$\begin{aligned}\sum_{k=1}^{\infty}|x_k+y_k|^p\omega_k &\le \sum_{k=1}^{\infty}|x_k+y_k|^{p-1}|x_k|\omega_k + \sum_{k=1}^{\infty}|x_k+y_k|^{p-1}|y_k|\omega_k\\ &\le \Big(\sum_{k=1}^{\infty}|x_k|^p\omega_k\Big)^{1/p}\Big(\sum_{k=1}^{\infty}|x_k+y_k|^{\overbrace{q(p-1)}^{p}}\omega_k\Big)^{1/q}\\ &\quad + \Big(\sum_{k=1}^{\infty}|y_k|^p\omega_k\Big)^{1/p}\Big(\sum_{k=1}^{\infty}|x_k+y_k|^{\overbrace{q(p-1)}^{p}}\omega_k\Big)^{1/q}\\ &= \Big(\sum_{k=1}^{\infty}|x_k+y_k|^p\omega_k\Big)^{1/q}\Big(\Big(\sum_{k=1}^{\infty}|x_k|^p\omega_k\Big)^{1/p} + \Big(\sum_{k=1}^{\infty}|y_k|^p\omega_k\Big)^{1/p}\Big).\end{aligned}$$

From this, we have

$$\begin{aligned}\Big(\sum_{k=1}^{\infty}|x_k+y_k|^p\omega_k\Big)^{\overbrace{1-1/q}^{1/p}} &\le \Big(\sum_{k=1}^{\infty}|x_k|^p\omega_k\Big)^{1/p} + \Big(\sum_{k=1}^{\infty}|y_k|^p\omega_k\Big)^{1/p},\\ \Big(\sum_{k=1}^{\infty}|x_k+y_k|^p\omega_k\Big)^{1/p} &\le \Big(\sum_{k=1}^{\infty}|x_k|^p\omega_k\Big)^{1/p} + \Big(\sum_{k=1}^{\infty}|y_k|^p\omega_k\Big)^{1/p}.\end{aligned}$$

We conclude that $\|\mathbf{x}+\mathbf{y}\|_{p,\omega} \le \|\mathbf{x}\|_{p,\omega} + \|\mathbf{y}\|_{p,\omega}$. □

Remark 1.1. By the previous theorem and the properties of real infinite series, we have that $\ell_{p,\omega}$ is a *normed* space.

Theorem 1.3. *The space $\ell_{p,\omega}$ is a Banach space for $1 \le p < +\infty$.*

Proof. We will only prove completeness, since the other properties of the norm are straightforward. Let $(\mathbf{x}_n)_{n\in\mathbb{N}}$ be Cauchy sequence in $\ell_{p,\omega}$ where $\mathbf{x}_n = (\mathbf{x}_1^{(n)}, \mathbf{x}_2^{(n)}, \mathbf{x}_3^{(n)}, \ldots)$. Then for any $\varepsilon > 0$ there exists $n_0 \in \mathbb{N}$ such that

$$\|\mathbf{x}_n - \mathbf{x}_m\|_{p,\omega} < \varepsilon \quad \text{for } n, m \ge n_0,$$

i. e.,

$$\Big(\sum_{k=1}^{\infty}|x_k^{(n)} - x_k^{(m)}|^p\omega_k\Big)^{1/p} < \varepsilon \tag{1.2}$$

whenever $n, m \geq n_0$. From (1.2) it is immediate that, for all $k = 1, 2, 3, \ldots$,

$$|x_k^{(n)} - x_k^{(m)}|\omega_k^{\frac{1}{p}} < \varepsilon \quad \text{for } n, m \geq n_0.$$

Note that

$$|x_k^{(n)}\omega_k^{\frac{1}{p}} - x_k^{(m)}\omega_k^{\frac{1}{p}}| = |x_k^{(n)} - x_k^{(m)}|\omega_k^{\frac{1}{p}} < \varepsilon \quad \text{for } n, m \geq n_0. \tag{1.3}$$

Fixing $k \in \mathbb{N}$, we have that $(x_k^{(1)}\omega_k^{\frac{1}{p}}, x_k^{(2)}\omega_k^{\frac{1}{p}}, x_k^{(3)}\omega_k^{\frac{1}{p}}, \ldots)$ is Cauchy sequence in $\mathbb{R}$, therefore convergent, so there exists $\tilde{x}_k\omega_k^{\frac{1}{p}} \in \mathbb{R}$ such that $\lim_{m\to\infty} x_k^{(m)}\omega_k^{\frac{1}{p}} = \tilde{x}_k\omega_k^{\frac{1}{p}}$ and $\lim_{m\to\infty} x_k^{(m)} = \tilde{x}_k$. Let us define $\tilde{\mathbf{x}} = (\tilde{x}_1, \tilde{x}_2, \tilde{x}_3, \ldots)$. We will show that $\tilde{\mathbf{x}} \in \ell_{p,\omega}$ and $\lim_{n\to\infty} \mathbf{x}_n = \tilde{\mathbf{x}}$. From (1.2),

$$\sum_{i=1}^{k} |x_i^{(n)} - x_i^{(m)}|^p \omega_i < \varepsilon^p,$$

and we conclude that

$$\sum_{i=1}^{k} |\tilde{x}_i - x_i^{(n)}|^p \omega_i = \sum_{i=1}^{k} \left| \lim_{m\to\infty} x_i^{(m)} - x_i^{(n)} \right|^p \omega_i < \varepsilon^p$$

whenever $n \geq n_0$. This shows that $\tilde{\mathbf{x}} - \mathbf{x}_n \in \ell_{p,\omega}$, and we also deduce that $\lim_{n\to+\infty} \mathbf{x}_n = \tilde{\mathbf{x}}$. Finally, by virtue of Theorem 1.2, we have

$$\begin{aligned}
\|\tilde{\mathbf{x}}\|_{p,\omega} = \left(\sum_{k=1}^{\infty} |\tilde{x}_k|^p \omega \right)^{\frac{1}{p}} &= \left(\sum_{k=1}^{\infty} |x_k^{(n)} + \tilde{x}_k - x_k^{(n)}|^p \omega \right)^{\frac{1}{p}} \\
&\leq \left(\sum_{k=1}^{\infty} |x_k^{(n)}|^p \omega \right)^{\frac{1}{p}} + \left(\sum_{k=1}^{\infty} |\tilde{x}_k - x_k^{(n)}|^p \omega \right)^{\frac{1}{p}} \\
&< \infty,
\end{aligned}$$

which shows that $\tilde{\mathbf{x}} \in \ell_{p,\omega}$. So we conclude that the Cauchy sequence $(\mathbf{x}_n)_{n\in\mathbb{N}}$ is convergent and the space $\ell_{p,\omega}$ is complete. □

Theorem 1.4. *The space $\ell_{p,\omega}$ is separable for $1 \leq p < \infty$.*

Proof. Define

$$M := \{\mathbf{q_n} : \mathbf{q_n} = (q_1, q_2, q_3, \ldots, q_n, 0, 0, 0, \ldots),\ n \in \mathbb{N},\ q_k \in \mathbb{Q}\}.$$

It is clear that $M \subsetneq \ell_{p,\omega}$. We will show that M is dense in $\ell_{p,\omega}$.

Let $\mathbf{x} = \{x_k\}_{k\in\mathbb{N}}$ be an arbitrary element of $\ell_{p,\omega}$. Then, for $\varepsilon > 0$, there exists $n \in \mathbb{N}$ (which depends on ε) such that

$$\sum_{k=n+1}^{\infty} |x_k|^p \omega_k < \frac{\varepsilon^p}{2}.$$

Now, since $\overline{\mathbb{Q}} = \mathbb{R}$, we have that for each $x_k \in \mathbb{R}$ there exists $q_k \in \mathbb{Q}$ such that

$$|x_k - q_k| < \frac{\varepsilon}{\sqrt[p]{2n\omega_k}}, \quad k = 1, 2, \ldots, n,$$

and then

$$\sum_{k=1}^{n} |x_k - q_k|^p \omega_k < \frac{\varepsilon^p}{2}.$$

Therefore, by setting $\mathbf{q} = (q_1, q_2, \ldots, q_n, 0, 0, \ldots)$

$$\|\mathbf{x} - \mathbf{q}\|_{p,\omega}^p = \sum_{k=1}^{n} |x_k - q_k|^p \omega_k + \sum_{k=n+1}^{\infty} |x_k|^p \omega_k < \frac{\varepsilon^p}{2} + \frac{\varepsilon^p}{2} < \varepsilon^p,$$

i. e., $\|\mathbf{x} - \mathbf{q}\|_{p,\omega} < \varepsilon$. This shows that M is dense in $\ell_{p,\omega}$ and, since M is countable as a countable union of countable sets, the space $\ell_{p,\omega}$ is separable. □

The next result tells us how to "compare" $\ell_{p,\omega}$ and $\ell_{q,\omega}$ spaces.

Theorem 1.5. *Let $1 \le p < q < \infty$. Then $\ell_{p,\omega} \subseteq \ell_{q,\omega}$ if and only if the weighted sequence ω satisfies the condition*

$$\omega_n \ge 1 \quad (n \in \mathbb{N}).$$

Proof. Let $1 \le p < q < \infty$ and $\ell_{p,\omega} \subseteq \ell_{q,\omega}$. Let $k \in \mathbb{N}$ be an arbitrary fixed natural number. Consider the following sequence $\mathbf{x} = (x_n)_{n\in\mathbb{N}}$ such that

$$x_n = \begin{cases} 1 & \text{if } k = n, \\ 0 & \text{if } k \neq n. \end{cases}$$

Then, since $\ell_{p,\omega} \subseteq \ell_{q,\omega}$, we have

$$\|\mathbf{x}\|_{p,\omega} \ge \|\mathbf{x}\|_{q,\omega}$$

and, because of the choice of the sequence $\mathbf{x}$,

$$\omega_k^{\frac{1}{p}} \ge \omega_k^{\frac{1}{q}},$$
$$\omega_k^{\frac{1}{p}-\frac{1}{q}} \ge 1.$$

Since $p < q$, one has $\frac{1}{p} - \frac{1}{q} > 0$. We conclude that $\omega_k \ge 1$ for an arbitrary $k \in \mathbb{N}$.

Now suppose that $\omega_n \geq 1$ is satisfied for all $n \in \mathbb{N}$, and let $\lambda > 1$ be such that $q = \lambda p$. Without loss of generality, assume that $\mathbf{x} \in \ell_{p,\omega}$ is such that $\sum_{n=1}^{\infty} |x_n|^p \omega_n = 1$. Then, for all $n \in \mathbb{N}$, we have that $|x_n|^p \omega_n \leq 1$, hence

$$\sum_{n=1}^{\infty} |x_n|^q \omega_n = \sum_{n=1}^{\infty} (|x_n|^p \omega_n)^{\lambda} \omega_n^{1-\lambda} \leq \sum_{n=1}^{\infty} (|x_n|^p \omega_n)^{\lambda} \leq \sum_{n=1}^{\infty} |x_n|^p \omega_n.$$

Thus, $\|\mathbf{x}\|_{p,\omega} \geq \|\mathbf{x}\|_{q,\omega}$, i. e., $\ell_{p,\omega} \subseteq \ell_{q,\omega}$. □

1.2 Strict convexity and uniform convexity

In the previous section, we proved that $\ell_{p,\omega}$ is a *vector* space, which implies that $t\mathbf{x} + (1-t)\mathbf{y} \in \ell_{p,\omega}$ for every $\mathbf{x}, \mathbf{y} \in \ell_{p,\omega}$ and $t \in [0,1]$. Thus the space $\ell_{p,\omega}$ is *convex* for $1 \leq p < \infty$. Moreover, we shall prove that the space $\ell_{p,\omega}$ has other "*convexity*" properties.

We recall now the concept of *strict convexity*.

Definition 1.4. A Banach space X is said to be *strictly convex* if, for $x, y \in X$ with $\|x\| = 1$, $\|y\| = 1$, and $x \neq y$, one has $\|\lambda x + (1-\lambda)y\| < 1$ for all $\lambda \in (0,1)$.

It is known that ℓ_p is a *strictly convex* space (see [50]). Also $\ell_{p,\omega}$ is *strictly convex* as we will show in the next theorem.

Theorem 1.6. *The space $\ell_{p,\omega}$ is strictly convex for $p \geq 1$.*

Proof. Let $\mathbf{x} = (x_n), \mathbf{y} = (y_n) \in \ell_{p,\omega}$ with $\|\mathbf{x}\|_{p,\omega} = 1$, $\|\mathbf{y}\|_{p,\omega} = 1$, and $p \geq 1$. Then $\|\mathbf{x}\omega^{\frac{1}{p}}\|_p = 1$ and $\|\mathbf{y}\omega^{\frac{1}{p}}\|_p = 1$. Since ℓ_p is *strictly convex* for $p \geq 1$, we have by characterization (1.1) that

$$\|\lambda\mathbf{x} + (1-\lambda)\mathbf{y}\|_{p,\omega} = \|\lambda\mathbf{x}\omega^{\frac{1}{p}} + (1-\lambda)\mathbf{y}\omega^{\frac{1}{p}}\|_p < 1.$$ □

We recall now the definition of a *uniformly convex* space.

Definition 1.5. A *Banach space* is said to be *uniformly convex* if for all $\varepsilon > 0$ there exists $\delta > 0$ such that the conditions $\|x\| \leq 1$, $\|y\| \leq 1$, and $\|x - y\| \geq \varepsilon$ imply $\|\frac{x+y}{2}\| \leq 1 - \delta$. The number

$$\delta(\varepsilon) = \inf\left\{1 - \left\|\frac{x+y}{2}\right\| : \|x\| = \|y\| = 1, \|x - y\| \geq \varepsilon\right\}$$

is called the *modulus of convexity*.

In order to prove the uniform convexity of $\ell_{p,\omega}$, we will need the following lemma. For its proof, see [16, Lemma 3.70, p. 112].

Lemma 1.1. *Let* $\mathbf{x}, \mathbf{y} \in \ell_p$ *and* $2 \leq p < \infty$, *then*

$$\|\mathbf{x}+\mathbf{y}\|_p^p + \|\mathbf{x}-\mathbf{y}\|_p^p \leq 2^{p-1}\left(\|\mathbf{x}\|_p^p + \|\mathbf{y}\|_p^p\right).$$

Let us show that the above result is also valid in the weighted context, i. e., it is not only valid for ℓ_p spaces, but also for $\ell_{p,\omega}$ spaces.

Theorem 1.7. *Let* $\mathbf{x}, \mathbf{y} \in \ell_{p,\omega}$ *and* $2 \leq p < \infty$, *then*

$$\|\mathbf{x}+\mathbf{y}\|_{p,\omega}^p + \|\mathbf{x}-\mathbf{y}\|_{p,\omega}^p \leq 2^{p-1}\left(\|\mathbf{x}\|_{p,\omega}^p + \|\mathbf{y}\|_{p,\omega}^p\right).$$

Proof. By Lemma 1.1, we have

$$\begin{aligned}\|\mathbf{x}+\mathbf{y}\|_{p,\omega}^p + \|\mathbf{x}-\mathbf{y}\|_{p,\omega}^p &= \left\|(\mathbf{x}+\mathbf{y})\omega^{\frac{1}{p}}\right\|_p^p + \left\|(\mathbf{x}-\mathbf{y})\omega^{\frac{1}{p}}\right\|_p^p \\ &= \left\|\mathbf{x}\omega^{\frac{1}{p}} + \mathbf{y}\omega^{\frac{1}{p}}\right\|_p^p + \left\|\mathbf{x}\omega^{\frac{1}{p}} - \mathbf{y}\omega^{\frac{1}{p}}\right\|_p^p \\ &\leq 2^{p-1}\left(\left\|\mathbf{x}\omega^{\frac{1}{p}}\right\|_p^p + \left\|\mathbf{y}\omega^{\frac{1}{p}}\right\|_p^p\right) \\ &= 2^{p-1}\left(\|\mathbf{x}\|_{p,\omega}^p + \|\mathbf{y}\|_{p,\omega}^p\right).\end{aligned}$$

□

Now, we are ready to prove the *uniform convexity* of $\ell_{p,\omega}$. First, we consider the case $2 \leq p \leq \infty$, as stated below.

Theorem 1.8. *The space* $\ell_{p,\omega}$ *is uniformly convex for* $2 \leq p < \infty$.

Proof. Let $\mathbf{x} = (x_n), \mathbf{y} = (y_n) \in \ell_{p,\omega}$ with $\|\mathbf{x}\|_{p,\omega} \leq 1$, $\|\mathbf{y}\|_{p,\omega} \leq 1$, and $\varepsilon > 0$ be such that $\|\mathbf{x}-\mathbf{y}\|_{p,\omega} \geq \varepsilon$. By Theorem 1.7, we have

$$\begin{aligned}\|\mathbf{x}+\mathbf{y}\|_{p,\omega}^p &\leq 2^{p-1}\left(\|\mathbf{x}\|_{p,\omega}^p + \|\mathbf{y}\|_{p,\omega}^p\right) - \|\mathbf{x}-\mathbf{y}\|_{p,\omega}^p \\ &\leq 2^{p-1}(2) - \varepsilon^p \\ &= 2^p\left(1 - \left(\frac{\varepsilon}{2}\right)^p\right).\end{aligned}$$

It follows that $\|\frac{\mathbf{x}+\mathbf{y}}{2}\|_{p,\omega}^p \leq 1 - (\frac{\varepsilon}{2})^p$ and so

$$\left\|\frac{\mathbf{x}+\mathbf{y}}{2}\right\|_{p,\omega}^p \leq 1 - \delta(\varepsilon),$$

where $\delta(\varepsilon) = 1 - (1 - (\frac{\varepsilon}{2})^p)$. □

In order to prove the *uniform convexity* of $\ell_{p,\omega}$ for the case $1 \leq p \leq 2$, we will need the following lemma (see [16, Lemma 3.70]).

Lemma 1.2. *Let* $\mathbf{x}, \mathbf{y} \in \ell_p$, $1 \le p \le 2$ *and* $q = \frac{p}{p-1}$, *then*

$$\|\mathbf{x}+\mathbf{y}\|_p^q + \|\mathbf{x}-\mathbf{y}\|_p^q \le 2\left(\|\mathbf{x}\|_p^p + \|\mathbf{y}\|_p^p\right)^{q-1}.$$

The previous result is also valid for the weighted case, as stated below.

Theorem 1.9. *Let* $\mathbf{x}, \mathbf{y} \in \ell_{p,\omega}$ *and* $1 < p \le 2$ *and* $q = \frac{p}{p-1}$, *then*

$$\|\mathbf{x}+\mathbf{y}\|_{p,\omega}^q + \|\mathbf{x}-\mathbf{y}\|_{p,\omega}^q \le 2\left(\|\mathbf{x}\|_{p,\omega}^p + \|\mathbf{y}\|_{p,\omega}^p\right)^{q-1}.$$

Proof. By Lemma 1.2, we have

$$\begin{aligned}
\|\mathbf{x}+\mathbf{y}\|_{p,\omega}^q + \|\mathbf{x}-\mathbf{y}\|_{p,\omega}^q &= \left\|(\mathbf{x}+\mathbf{y})\omega^{\frac{1}{p}}\right\|_p^q + \left\|(\mathbf{x}-\mathbf{y})\omega^{\frac{1}{p}}\right\|_p^q \\
&= \left\|\mathbf{x}\omega^{\frac{1}{p}} + \mathbf{y}\omega^{\frac{1}{p}}\right\|_p^q + \left\|\mathbf{x}\omega^{\frac{1}{p}} - \mathbf{y}\omega^{\frac{1}{p}}\right\|_p^q \\
&\le 2\left(\left\|\mathbf{x}\omega^{\frac{1}{p}}\right\|_p^p + \left\|\mathbf{y}\omega^{\frac{1}{p}}\right\|_p^p\right)^{q-1} \\
&\le 2\left(\|\mathbf{x}\|_{p,\omega}^p + \|\mathbf{y}\|_{p,\omega}^p\right)^{q-1}.
\end{aligned}$$

□

We are now ready to prove the *uniform convexity* of $\ell_{p,\omega}$ for the remaining case $1 < p \le 2$.

Theorem 1.10. *The space* $\ell_{p,\omega}$ *is uniformly convex for* $1 < p \le 2$.

Proof. Let $\mathbf{x} = (x_n), \mathbf{y} = (y_n) \in \ell_{p,\omega}$ with $\|\mathbf{x}\|_{p,\omega} \le 1$, $\|\mathbf{y}\|_{p,\omega} \le 1$, and $\varepsilon > 0$ be such that $\|\mathbf{x}-\mathbf{y}\|_{p,\omega} \ge \varepsilon$. By Theorem 1.9,

$$\begin{aligned}
\|\mathbf{x}+\mathbf{y}\|_{p,\omega}^q &\le 2\left(\|\mathbf{x}\|_{p,\omega}^p + \|\mathbf{y}\|_{p,\omega}^p\right)^{q-1} - \|\mathbf{x}-\mathbf{y}\|_{p,\omega}^q \\
&\le 2(2)^{q-1} - \varepsilon^q \\
&= 2^q\left(1 - \left(\frac{\varepsilon}{2}\right)^q\right).
\end{aligned}$$

Hence, we have

$$\left\|\frac{\mathbf{x}+\mathbf{y}}{2}\right\|_{p,\omega} \le \left(1 - \left(\frac{\varepsilon}{2}\right)^q\right)^{\frac{1}{q}},$$

thus

$$\left\|\frac{\mathbf{x}+\mathbf{y}}{2}\right\|_{p,\omega} \le 1 - \delta(\varepsilon),$$

where $\delta(\varepsilon) = 1 - (1 - (\frac{\varepsilon}{2})^q)^{\frac{1}{q}}$, as required. □

1.3 Dual space of $\ell_{p,\omega}$

We study now the dual space of $\ell_{p,\omega}$, i. e., the space of all linear bounded functionals on $\ell_{p,\omega}$. As we have shown in the last section, it seems that properties of ℓ_p are also valid in $\ell_{p,\omega}$ (of course, in the weighted context). Also, remember that the dual space of ℓ_p is ℓ_q, i. e., $(\ell_p)' = \ell_q$ where $1 < p < \infty$ and $1 < q$ are such that $\frac{1}{p} + \frac{1}{q} = 1$ (see [10, 16, 50]). Then, in the weighted context one has $(\ell_{p,\omega})' = \ell_{q,\omega}$, as expected.

We begin by stating and proving the *Riesz representation theorem* for $\ell_{p,\omega}$ spaces.

Definition 1.6. Define the sequence $\mathbf{e}^{(\mathbf{n})} = \{e_k^{(n)}\}_{k\in\mathbb{N}}$ as follows:

$$e_k^{(n)} = \begin{cases} 1 & \text{if } k = n, \\ 0 & \text{if } k \neq n. \end{cases}$$

We claim that $\mathbf{e}^{(\mathbf{n})} \in \ell_{p,\omega}$ for all weight sequences ω, and $\mathbf{e}^{(\mathbf{n})} \cdot \mathbf{x} \in \ell_{p,\omega}$ for all $\mathbf{x} \in \ell_{p,\omega}$. In fact,

$$\sum_{k=1}^{\infty} |e_k^{(n)}|^p \omega_k = |e_n^{(n)}|^p \omega_n = 1^p \omega_n = \omega_n < \infty.$$

Therefore, $\mathbf{e}^{(\mathbf{n})} \in \ell_{p,\omega}$ and $\mathbf{e}^{(\mathbf{n})} \cdot \mathbf{x} \in \ell_{p,\omega}$ for all $\mathbf{x} \in \ell_{p,\omega}$, since $\ell_{p,\omega}$ is a Banach space. Also,

$$\|\mathbf{e}^{(\mathbf{n})}\|_{p,\omega} = \left(\sum_{k=1}^{\infty} |e_k^{(n)}|^p \omega_k \right)^{1/p} = (\omega_n)^{1/p}. \tag{1.4}$$

We are ready to prove the main result of this section.

Theorem 1.11 (Riesz representation theorem for the $\ell_{p,\omega}$ space). *Let $1 < p < \infty$ and $1 < q < \infty$ be such that $\frac{1}{p} + \frac{1}{q} = 1$. For every bounded linear functional $f \in (\ell_{p,\omega})'$, there exists $\mathbf{z} \in \ell_{q,\omega}$ such that*

$$f(\mathbf{x}) = \sum_{k=1}^{\infty} x_k z_k \omega_k$$

for all $\mathbf{x} \in \ell_{p,\omega}$ and $\|\mathbf{z}\|_{q,\omega} = \|f\|$.

Proof. Let $\mathbf{x} = (x_n)_{n\in\mathbb{N}} \in \ell_{p,\omega}$. We have shown in (1.4) that $\mathbf{e}^{(\mathbf{n})} \in \ell_{p,\omega}$, thus

$$\lim_{n\to\infty} \left\| \mathbf{x} - \sum_{k=1}^{n} x_k \mathbf{e}^{(\mathbf{k})} \right\|_{p,\omega}^p = \lim_{n\to\infty} \sum_{k=n+1}^{\infty} |x_k|^p \omega_k = 0.$$

This shows that each $\mathbf{x} \in \ell_{p,\omega}$ has a unique representation of the form

$$\mathbf{x} = \sum_{k=1}^{\infty} x_k \mathbf{e}^{(\mathbf{k})},$$

which means that $\{\mathbf{e}^{(\mathbf{k})}\}_{k\in\mathbb{N}}$ is a Schauder basis of $\ell_{p,\omega}$.

Now, let $f \in (\ell_{p,\omega})'$ and define the sequence $\mathbf{z} = (z_k)_{k\in\mathbb{N}}$ as follows:

$$z_k = \frac{f(\mathbf{e}^{(\mathbf{k})})}{\omega_k}.$$

Given that f is linear and bounded, it is continuous, and for each $\mathbf{x} \in \ell_{p,\omega}$ we have

$$\begin{aligned} f(\mathbf{x}) &= f\left(\sum_{k=1}^{\infty} x_k \mathbf{e}^{(\mathbf{k})}\right) \\ &= f\left(\lim_{n\to\infty} \sum_{k=1}^{n} x_k \mathbf{e}^{(\mathbf{k})}\right) \\ &= \lim_{n\to\infty} f\left(\sum_{k=1}^{n} x_k \mathbf{e}^{(\mathbf{k})}\right) \\ &= \lim_{n\to\infty} \sum_{k=1}^{n} x_k f(\mathbf{e}^{(\mathbf{k})}) \\ &= \lim_{n\to\infty} \sum_{k=1}^{n} x_k z_k \omega_k = \sum_{k=1}^{\infty} x_k z_k \omega_k. \end{aligned}$$

We claim that $\mathbf{z} \in \ell_{q,\omega}$. Indeed, consider the sequence $\zeta^{(n)} = (\zeta_k^{(n)})_{k\in\mathbb{N}}$ defined as

$$\zeta_k^{(n)} = \begin{cases} \frac{|z_k|^q}{z_k} & \text{if } k \le n \text{ and } z_k \ne 0, \\ 0 & \text{if } n < k \text{ or } z_k = 0. \end{cases}$$

Then $\zeta^{(n)} \in \ell_{p,\omega}$ (since it has a finite number of nonzero elements).

So

$$f(\zeta^{(n)}) = \sum_{k=1}^{\infty} \zeta_k^{(n)} z_k \omega_k = \sum_{k=1}^{n} |z_k|^q \omega_k$$

and

$$\begin{aligned} |f(\zeta^{(n)})| &\le \|f\| \|\zeta^{(n)}\|_{p.\omega} \\ &= \|f\| \left(\sum_{k=1}^{\infty} |\zeta_k^{(n)}|^p \omega_k\right)^{1/p} \end{aligned}$$

$$= \|f\| \left(\sum_{k=1}^{n} |z_k|^{(q-1)p} \omega_k \right)^{1/p}$$
$$= \|f\| \left(\sum_{k=1}^{n} |z_k|^{q} \omega_k \right)^{1/p},$$

hence

$$|f(\zeta^{(n)})| = \sum_{k=1}^{n} |z_k|^q \omega_k \le \|f\| \left(\sum_{k=1}^{n} |z_k|^q \omega_k \right)^{1/p},$$
$$\left(\sum_{k=1}^{n} |z_k|^q \omega_k \right)^{1-1/p} \le \|f\|,$$
$$\left(\sum_{k=1}^{n} |z_k|^q \omega_k \right)^{1/q} \le \|f\|.$$

Taking the limit as $n \to \infty$, we conclude that

$$\|\mathbf{z}\|_{q,\omega} \le \|f\|, \tag{1.5}$$

and so $\mathbf{z} \in \ell_{q,\omega}$.

It remains to show that $\|\mathbf{z}\|_{q,\omega} = \|f\|$. Since f is linear and

$$|f(x)| = \left| \sum_{k=1}^{\infty} x_k z_k \omega_k \right|$$
$$\le \sum_{k=1}^{\infty} |x_k z_k| \omega_k$$
$$\le \|\mathbf{x}\|_{p,\omega} \|\mathbf{z}\|_{q,\omega},$$

by the weighted Hölder inequality (Theorem 1.1), for $\mathbf{x} \neq \mathbf{0}$ one has

$$\frac{|f(\mathbf{x})|}{\|\mathbf{x}\|_{p,\omega}} \le \|\mathbf{z}\|_{q,\omega},$$

and so

$$\|f\| \le \|\mathbf{z}\|_{q,\omega}.$$

Combining the above inequality and (1.5), we conclude that

$$\|\mathbf{z}\|_{q,\omega} = \|f\|. \qquad \square$$

2 A few weighted inequalities

The Hardy inequality holds a fundamental place in the realm of mathematical inequalities, particularly serving as a generalization of many classical results. Historically, Hilbert inequality, discovered in the early 1900s, played a pivotal role in the development of the Hardy inequality, with Hilbert's work motivating Hardy to extend and deepen the understanding of such inequalities (see [51]). Furthermore, Carlson inequality, initially regarded as a limiting case not covered by the Hölder–Rogers inequality (see [53]), gained additional importance when Hardy presented a proof using merely the Schwarz inequality [44].

This connection between Hardy, Hilbert, and Carlson inequalities showcases how these foundational results not only interrelate but also lay the groundwork for more general and robust versions, influencing a wide range of mathematical fields. In this chapter, we will establish the Hardy inequality in its weighted form, along with a related weighted inequality. Additionally, we will explore the connections between the Hardy, Hilbert, and Carlson inequalities, highlighting their interrelations and significance in the broader context of inequality theory.

2.1 Hardy's weighted inequality

In order to prove the next theorem, we will need the following well-known inequality. For its proof, see [10, 16].

Lemma 2.1 (Young's inequality). *Let $a, b \geq 0$ and $p, q > 1$ be such that $\frac{1}{p} + \frac{1}{q} = 1$. Then*

$$ab \leq \frac{a^p}{p} + \frac{b^q}{q}.$$

The following theorem (nonweighted version) may be found in [16, Theorem 2.16, p. 35]. We state and prove the *Hardy's weighted inequality*.

Theorem 2.1 (Hardy's weighted inequality). *Let $\{a_n\}_{n\in\mathbb{N}}$ be a nonnegative sequence of real numbers such that $\sum_{n=1}^{\infty} a_n^p \omega_n < \infty$ where $\omega_1 \geq \omega_2 \geq \cdots$, and $\omega_n > 0$ for all $n \in \mathbb{N}$. Then*

$$\sum_{n=1}^{\infty} \left(\frac{1}{n} \sum_{k=1}^{n} a_k \right)^p \omega_n \leq \left(\frac{p}{p-1} \right)^p \sum_{n=1}^{\infty} a_n^p \omega_n.$$

Proof. Let $\alpha_n = \frac{A_n}{n}$ where $A_n = a_1 + a_2 + \cdots + a_n$, i. e., $A_n = n\alpha_n$, then

$$a_1 + a_2 + \cdots + a_n = n\alpha_n,$$

https://doi.org/10.1515/9783112223246-002

from which we have that

$$na_n - (n-1)a_{n-1} = a_n,$$

and so

$$a_n\omega_n = na_n\omega_n - (n-1)a_{n-1}\omega_n \quad \text{for } n \in \mathbb{N}.$$

Now, let us consider

$$\begin{aligned} a_n^p\omega_n - \frac{p}{p-1}a_n^{p-1}a_n\omega_n &= a_n^p\omega_n - \frac{p}{p-1}a_n^{p-1}(na_n - (n-1)a_{n-1})\omega_n \\ &= a_n^p\omega_n - \frac{p}{p-1}a_n^{p-1}na_n\omega_n + \frac{p}{p-1}a_n^{p-1}(n-1)a_{n-1}\omega_n \\ &= a_n^p\omega_n - \frac{p}{p-1}na_n^p\omega_n + \frac{p(n-1)}{p-1}a_n^{p-1}a_{n-1}\omega_n, \end{aligned}$$

for $p > 1$ and $n \in \mathbb{N}$.

Letting $q > 1$ be such that $\frac{1}{p} + \frac{1}{q} = 1$ and by *Young's inequality* stated in Lemma 2.1, we have

$$\begin{aligned} \frac{p(n-1)}{p-1}a_n^{p-1}a_{n-1}\omega_n &= \frac{p(n-1)}{p-1}a_{n-1}\omega_n^{1/p}a_n^{p-1}\omega_n^{1/q} \\ &\le \frac{p(n-1)}{p-1}\left(\frac{a_{n-1}^p\omega_n^{(1/p)p}}{p} + \frac{a_n^{(p-1)q}\omega_n^{(1/q)q}}{q}\right) \\ &= \frac{p(n-1)}{p-1}\frac{a_{n-1}^p\omega_n^{(1/p)p}}{p} + \frac{p(n-1)}{p-1}\frac{a_n^{(p-1)q}\omega_n^{(1/q)q}}{q} \\ &= \frac{(n-1)}{p-1}a_{n-1}^p\omega_n + (n-1)a_n^p\omega_n. \end{aligned}$$

Hence

$$\begin{aligned} &a_n^p\omega_n - \frac{p}{p-1}a_n^{p-1}a_n\omega_n \\ &\quad\le a_n^p\omega_n - \frac{p}{p-1}na_n^p\omega_n + \frac{(n-1)}{p-1}a_{n-1}^p\omega_n + (n-1)a_n^p\omega_n \\ &\quad= \frac{pa_n^p\omega_n - a_n^p\omega_n - pna_n^p\omega_n}{p-1} + \frac{(n-1)a_{n-1}^p\omega_n + (p-1)(n-1)a_n^p\omega_n}{p-1} \\ &\quad= \frac{pa_n^p\omega_n - a_n^p\omega_n - pna_n^p\omega_n + (n-1)a_{n-1}^p\omega_n + (pn-p-n+1)a_n^p\omega_n}{p-1} \\ &\quad= \frac{1}{p-1}\left[(n-1)a_{n-1}^p\omega_n - na_n^p\omega_n\right]. \end{aligned}$$

Then for a fixed $N \in \mathbb{N}$, we have

$$\begin{aligned}\sum_{n=1}^{N} a_n^p \omega_n - \frac{p}{p-1}\sum_{n=1}^{N} a_n^{p-1} a_n \omega_n &\le \frac{1}{p-1}\sum_{n=1}^{N}(n-1)a_{n-1}^p \omega_n - n a_n^p \omega_n \\ &= \frac{1}{p-1}(-a_1^p\omega_1 + (a_1^p - 2a_2^p)\omega_2 + (2a_2^p - 3a_3^p)\omega_3 \\ &\quad + \dots + ((N-1)a_{N-1}^p - Na_N^p)\omega_N) \\ &= \frac{1}{p-1}(-a_1^p\omega_1 + a_1^p\omega_2 - 2a_2^p\omega_2 + 2a_2^p\omega_3 - 3a_3^p\omega_3 \\ &\quad + \dots + (N-1)a_{N-1}^p\omega_N - Na_N^p\omega_N).\end{aligned}$$

Note that $-ka_k^p\omega_k + ka_k^p\omega_{k+1} < 0$ for $k = 1, 2, \dots, N$, since $(\omega_n)_{n\in\mathbb{N}}$ is a decreasing positive sequence. Hence

$$\sum_{n=1}^{N} a_n^p \omega_n - \frac{p}{p-1}\sum_{n=1}^{N} a_n^{p-1} a_n \omega_n \le 0,$$

and so

$$\sum_{n=1}^{N} a_n^p \omega_n \le \frac{p}{p-1}\sum_{n=1}^{N} a_n^{p-1} a_n \omega_n.$$

Taking the limit as $N \to \infty$, we obtain

$$\sum_{n=1}^{\infty} a_n^p \omega_n \le \frac{p}{p-1}\sum_{n=1}^{\infty} a_n^{p-1} a_n \omega_n$$

and, applying the *weighted Hölder's inequality* (Theorem 1.1), we have

$$\begin{aligned}\sum_{n=1}^{\infty} a_n^p \omega_n &\le \frac{p}{p-1}\left(\sum_{n=1}^{\infty} a_n^p \omega_n\right)^{1/p}\left(\sum_{n=1}^{\infty} a_n^{(p-1)q} \omega_n\right)^{1/q} \\ &\le \frac{p}{p-1}\left(\sum_{n=1}^{\infty} a_n^p \omega_n\right)^{1/p}\left(\sum_{n=1}^{\infty} a_n^p \omega_n\right)^{1/q},\end{aligned}$$

and so

$$\left(\sum_{n=1}^{\infty} a_n^p \omega_n\right)^{1/p} \le \frac{p}{p-1}\left(\sum_{n=1}^{\infty} a_n^p \omega_n\right)^{1/p}.$$

Then we conclude that

$$\sum_{n=1}^{\infty}\left(\frac{1}{n}\sum_{k=1}^{n} a_k\right)^p \omega_n \le \left(\frac{p}{p-1}\right)^p \sum_{n=1}^{\infty} a_n^p \omega_n.$$

□

The following inequality is due to the Swedish mathematician *Torsten Carleman* [5], and it is known as *Carleman's inequality*.

Theorem 2.2 (Carleman's inequality). *Let $\{a_n\}_{n\in\mathbb{N}}$ be a positive sequence. Then*

$$\sum_{n=1}^{\infty}\left(\sqrt[n]{a_1 a_2 \cdots a_n}\right) \le e \sum_{n=1}^{\infty} a_n.$$

Its weighted version is also true.

Theorem 2.3 (Carleman's weighted inequality). *Let $\{a_n\}_{n\in\mathbb{N}}$ be a positive sequence and $\{\omega_n\}_{n\in\mathbb{N}}$ a positive decreasing sequence. Then*

$$\sum_{n=1}^{\infty}\left(\sqrt[n]{a_1 a_2 \cdots a_n}\right)\omega_n \le e \sum_{n=1}^{\infty} a_n \omega_n.$$

Proof. By Theorem 2.1,

$$\sum_{n=1}^{\infty}\left(\frac{a_1 + a_2 + \cdots + a_n}{n}\right)^p \omega_n \le \left(\frac{p}{p-1}\right)^p \sum_{n=1}^{\infty} a_n^p \omega_n,$$

from which we have

$$\sum_{n=1}^{\infty}\left(\frac{a_1^{1/p} + a_2^{1/p} + \cdots + a_n^{1/p}}{n}\right)^p \omega_n \le \left(\frac{p}{p-1}\right)^p \sum_{n=1}^{\infty} a_n \omega_n.$$

But, by the geometric–arithmetic means' inequality, we know that

$$\left(\sqrt[n]{a_1^{1/p} a_2^{1/p} \cdots a_n^{1/p}}\right)^p \le \left(\frac{a_1^{1/p} + a_2^{1/p} + \cdots + a_n^{1/p}}{n}\right)^p.$$

Hence

$$\begin{aligned}\sum_{n=1}^{\infty}\left(\sqrt[n]{a_1 a_2 \cdots a_n}\right)\omega_n &\le \lim_{p\to\infty}\left(1 + \frac{1}{p-1}\right)^{p-1} \lim_{p\to\infty}\left(1 + \frac{1}{p-1}\right) \sum_{n=1}^{\infty} a_n \omega_n \\ &\le e \sum_{n=1}^{\infty} a_n \omega_n,\end{aligned}$$

as we wanted. □

2.2 Carlson's inequality

The next inequality is named after the Swedish professor *Fritz Carlson* (see [6]). He proved that, for any sequence $\{a_n\}_{n\in\mathbb{N}}$ of nonnegative real numbers, not all zero, the following inequality holds:

$$\left(\sum_{n=1}^{\infty} a_n\right)^4 < \pi^2\left(\sum_{n=1}^{\infty} a_n^2\right)\left(\sum_{n=1}^{\infty} n^2 a_n^2\right). \tag{2.1}$$

It seems that *Carlson* thought he had discovered an inequality which is independent of other inequalities, viz. the *Hölder's* inequality.

G. H. Hardy [44], who in 1936 provided two straightforward proofs of the Carlson's inequality (2.1), likely had the most significant influence on subsequent work in this area. One of his proofs proved to be remarkably powerful, and the fundamental simplicity of his approach was later adopted by numerous mathematicians.

We present now the Hardy's proof of Carlson's inequality. For more on the Carlson's inequality, the reader is invited to check [52] and the references therein.

Theorem 2.4 (Carlson's inequality). *Let $\{a_n\}_{n\in\mathbb{N}}$ be a sequence of real numbers, not all zero. Then*

$$\sum_{n=1}^{\infty} a_n \le \sqrt{\pi}\left(\sum_{n=1}^{\infty} a_n^2\right)^{\frac{1}{4}}\left(\sum_{n=1}^{\infty} n^2 a_n^2\right)^{\frac{1}{4}}.$$

Proof. Letting $\alpha, \beta > 0$ and applying the *Cauchy–Schwarz inequality*, we have

$$\begin{aligned}\left(\sum_{n=1}^{\infty} a_n\right)^2 &= \left(\sum_{n=1}^{\infty} a_n\sqrt{\alpha+\beta n^2}\frac{1}{\sqrt{\alpha+\beta n^2}}\right)^2\\ &\le \sum_{n=1}^{\infty} a_n^2(\alpha+\beta n^2)\sum_{n=1}^{\infty}\frac{1}{\alpha+\beta n^2}.\end{aligned}$$

Now, observe that

$$\begin{aligned}\sum_{n=1}^{\infty}\frac{1}{\alpha+\beta n^2} &\le \int_0^{\infty}\frac{dx}{\alpha+\beta x^2}\\ &\le \frac{1}{\sqrt{\alpha\beta}}\arctan\left(\sqrt{\frac{\beta}{\alpha}}x\right)\Bigg|_0^{\infty}\\ &= \frac{\pi}{2\sqrt{\alpha\beta}}.\end{aligned}$$

Hence

$$\left(\sum_{n=1}^{\infty} a_n\right)^2 \le \frac{\pi}{2\sqrt{\alpha\beta}} \sum_{n=1}^{\infty} a_n^2(\alpha + \beta n^2)$$
$$= \frac{\pi}{2\sqrt{\alpha\beta}}\left(\alpha \sum_{n=1}^{\infty} a_n^2 + \beta \sum_{n=1}^{\infty} n^2 a_n^2\right).$$

Let us take $S = \sum_{n=1}^{\infty} a_n^2$ and $T = \sum_{n=1}^{\infty} n^2 a_n^2$, together with $\alpha = T$ and $\beta = S$. Then

$$\frac{\pi}{2\sqrt{\alpha\beta}}\left(\alpha \sum_{n=1}^{\infty} a_n^2 + \beta \sum_{n=1}^{\infty} n^2 a_n^2\right) = \frac{\pi}{2\sqrt{\alpha\beta}}(\alpha S + \beta T)$$
$$= \frac{\pi}{2\sqrt{TS}}(TS + ST)$$
$$= \pi\sqrt{ST}.$$

Finally,

$$\left(\sum_{n=1}^{\infty} a_n\right)^2 \le \pi\sqrt{ST}$$
$$= \pi\sqrt{\sum_{n=1}^{\infty} a_n^2}\sqrt{\sum_{n=1}^{\infty} n^2 a_n^2}$$
$$= \pi\left(\sum_{n=1}^{\infty} a_n^2\right)^{\frac{1}{2}}\left(\sum_{n=1}^{\infty} n^2 a_n^2\right)^{\frac{1}{2}}.$$

Therefore,

$$\sum_{n=1}^{\infty} a_n \le \sqrt{\pi}\left(\sum_{n=1}^{\infty} a_n^2\right)^{\frac{1}{4}}\left(\sum_{n=1}^{\infty} n^2 a_n^2\right)^{\frac{1}{4}}.$$ □

Let us recall now the *Hilbert inequality*.

Theorem 2.5 (Hilbert inequality). *Let $\{a_n\}_{n\in\mathbb{N}}$, $\{b_n\}_{n\in\mathbb{N}}$ be sequences of nonnegative real numbers such that $\sum_{n=1}^{\infty} a_n^2$ and $\sum_{n=1}^{\infty} b_n^2$ are convergent. Then*

$$\sum_{n=1}^{\infty}\sum_{m=1}^{\infty} \frac{a_n b_m}{n+m} \le \pi\left(\sum_{n=1}^{\infty} a_n^2\right)^{\frac{1}{2}}\left(\sum_{m=1}^{\infty} b_m^2\right)^{\frac{1}{2}}. \tag{2.2}$$

A proof of the above inequality may be found in [16]. For a historical background, see [41, 42, 46].

We prove now that *Carlson's inequality* may be obtained from *Hilbert inequality*. In fact, (2.2) implies that

$$
\begin{aligned}
\left(\sum_{n=1}^{\infty} a_n\right)^2 &= \sum_{n=1}^{\infty} a_n \sum_{m=1}^{\infty} a_m \\
&= \sum_{n=1}^{\infty}\sum_{m=1}^{\infty} \frac{n+m}{n+m} a_n a_m \\
&= \sum_{n=1}^{\infty}\sum_{m=1}^{\infty} \frac{n a_n a_m}{n+m} + \sum_{n=1}^{\infty}\sum_{m=1}^{\infty} \frac{m a_n a_m}{n+m} \\
&\le \pi\left(\sum_{n=1}^{\infty}(n a_n)^2\right)^{\frac{1}{2}}\left(\sum_{m=1}^{\infty} a_m^2\right)^{\frac{1}{2}} + \pi\left(\sum_{m=1}^{\infty}(m a_m)^2\right)^{\frac{1}{2}}\left(\sum_{n=1}^{\infty} a_n^2\right)^{\frac{1}{2}} \\
&= \pi\left(\sum_{n=1}^{\infty}(n a_n)^2\right)^{\frac{1}{2}}\left(\sum_{n=1}^{\infty} a_n^2\right)^{\frac{1}{2}} + \pi\left(\sum_{n=1}^{\infty}(n a_n)^2\right)^{\frac{1}{2}}\left(\sum_{n=1}^{\infty} a_n^2\right)^{\frac{1}{2}} \\
&= 2\pi\left(\sum_{n=1}^{\infty}(n a_n)^2\right)^{\frac{1}{2}}\left(\sum_{n=1}^{\infty} a_n^2\right)^{\frac{1}{2}},
\end{aligned}
$$

hence

$$
\sum_{n=1}^{\infty} a_n \le \sqrt{2\pi}\left(\sum_{n=1}^{\infty}(n a_n)^2\right)^{\frac{1}{4}}\left(\sum_{n=1}^{\infty} a_n^2\right)^{\frac{1}{4}}.
$$

3 Weighted L_p spaces

We now arrive at the central theme of this book: the study of *weighted Lebesgue spaces*. These spaces, which generalize the classical L_p spaces by incorporating weight functions, play a fundamental role in modern functional analysis, harmonic analysis, and partial differential equations. By modifying the underlying measure with a weight, we obtain refined function spaces that capture subtle integrability and growth properties essential for applications in areas such as interpolation theory, Sobolev embeddings, and singular integral operators.

In a similar way as we have done in Chapter 1, we investigate key properties of weighted Lebesgue spaces, including completeness, separability, duality, and also relevant inequalities such as those of Hölder and Minkowski. Additionally, we study geometric aspects such as strict and uniform convexity, which have significant implications for optimization and variational problems. A particular focus is given to the Hardy operator, a classical tool in analysis.

Let us begin by defining weight functions.

Definition 3.1. A weight ω is a nonnegative locally integrable function on $(0, +\infty)$ that takes values in $(0, +\infty)$ almost everywhere. Therefore, weights are allowed to be zero or infinite only on a set of Lebesgue measure zero.

Remark 3.1. From now on, unless stated otherwise, the symbol ω denotes a weight.

According to Definition 3.1, if ω is a weight and $\frac{1}{\omega}$ is locally integrable, then $\frac{1}{\omega}$ is also a weight.

Given a weight ω and a measurable set E, we use the notation

$$\omega(E) = \int_E \omega(x)\,dx$$

to denote the ω-measure of the set E.

Since weights are locally integrable functions, $\omega(E) < +\infty$ for all sets E contained in some ball.

We have come to the definition of the main set of this book, i. e., weighted Lebesgue spaces.

Definition 3.2. The weighted L_p space is denoted by $L_p(\omega)$, for $1 \le p < +\infty$. It consists of all measurable functions f such that

$$\|f\|_{L_p(\omega)} := \left(\int_0^{+\infty} |f(x)|^p\,\omega(x)\,dx\right)^{\frac{1}{p}} < +\infty.$$

That is,

https://doi.org/10.1515/9783112223246-003

$$L_p(\omega) := \left\{ f : \|f\|_{L_p(\omega)} := \left(\int_0^{+\infty} |f(x)|^p \omega(x)\, dx \right)^{\frac{1}{p}} < +\infty \right\}.$$

Also, we denote

$$(L_p(\omega))^d := \{f \in L_p(\omega) : f \text{ is decreasing}\}.$$

It is not difficult to verify that $L_p(\omega)$ with $1 \le p < +\infty$ is a vector space. In fact, note that if $f, g \in L_p(\omega)$, then by the inequality

$$|f + g|^p \le (|f| + |g|)^p \le (2\max\{|f|, |g|\})^p = 2^p \max\{|f|^p, |g|^p\} \le 2^p(|f|^p + |g|^p) \tag{3.1}$$

we have that $f + g \in L_p(\omega)$. By means of the convexity of $\varphi(t) = t^p$ for $p \ge 1$, one proves that, in fact,

$$|f + g|^p \le 2^{p-1}(|f|^p + |g|^p). \tag{3.2}$$

Moreover, if $f \in L_p(\omega)$ and $\alpha \in \mathbb{R}$, then $\alpha f \in L_p(\omega)$. Also, recalling that the positive part f^+ of a function f is defined as $f^+(x) = \max\{0, f(x)\}$, and similarly the negative part f^- of f is defined as $f^-(x) = \max\{0, -f(x)\}$, the inequalities $0 \le f^+ \le |f|$, $0 \le f^- \le |f|$ imply that f^+ and f^- belong to $L_p(\omega)$.

Classical inequalities in the L_p spaces, such as those of Hölder and Minkowski, hold in the weighted context. Let us begin with the former.

Theorem 3.1 (Hölder's inequality). *Let p and q be extended nonnegative numbers such that $\frac{1}{p} + \frac{1}{q} = 1$ and $f \in L_p(\omega)$, $g \in L_p(\omega)$. Then $fg \in L_1(\omega)$ and*

$$\int_0^{+\infty} |f(x)g(x)|\, \omega(x)\, dx \le \|f\|_{L_p(\omega)} \|g\|_{L_q(\omega)}.$$

The equality holds if there are constants A and B, not simultaneously zero, such that $A\,|f|^p = B\,|g|^q$ ω-a. e.

Proof. Let $f \in L_p(\omega)$ and $g \in L_p(\omega)$, then

$$\begin{aligned}
\int_0^{+\infty} |f(x)g(x)|\, \omega(x)\, dx &= \int_0^{+\infty} |f(x)|\, \omega^{\frac{1}{p}}(x)\, |g(x)|\, \omega^{\frac{1}{q}}(x)\, dx \\
&\le \left(\int_0^{+\infty} |f(x)|^p\, \omega(x)\, dx \right)^{\frac{1}{p}} \left(\int_0^{+\infty} |g(x)|^p \quad \omega(x)\, dx \right)^{\frac{1}{q}} \\
&\le \|f\|_{L_p(\omega)} \|g\|_{L_q(\omega)}.
\end{aligned}$$

Finally, choosing $A = \|g\|^q_{L_q(\omega)}$ and $B = \|f\|^p_{L_p(\omega)}$ such that $A\,|f|^p = B\,|g|^q$, we have

$$|f(x)| = \|f\|_{L_p(\omega)} \frac{|g(x)|^{\frac{q}{p}}}{\|g\|_{L_q(\omega)}^{\frac{q}{p}}}$$

and by integrating get

$$\int_0^{+\infty} |f(x)g(x)|\, \omega(x)\, dx = \|f\|_{L_p(\omega)} \|g\|_{L_q(\omega)}. \qquad \square$$

Theorem 3.2 (Minkowski inequality). *Let $1 \le p < +\infty$ and $f, g \in L_p(\omega)$. Then*

$$\|f + g\|_{L_p(\omega)} \le \|f\|_{L_p(\omega)} + \|g\|_{L_p(\omega)}.$$

The equality holds if $A\,|f| = B\,|g|$ ω-a. e. for A and B of the same sign and not simultaneously zero.

Proof. Let us check the equality. Let A and B be numbers of the same sign and not simultaneously zero such that $A\,|f| = B\,|g|$ ω-a. e. Then $A\,\|f\|_{L_p(\omega)} = B\,\|g\|_{L_p(\omega)}$, i. e.,

$$\|f\|_{L_p(\omega)} = \frac{B}{A}\|g\|_{L_p(\omega)}.$$

Moreover,

$$\begin{aligned} \|f + g\|_{L_p(\omega)} = \left\|\frac{B}{A}g + g\right\|_{L_p(\omega)} &= \frac{B + A}{A}\,\|g\|_{L_p(\omega)} \\ &= \frac{B}{A}\,\|g\|_{L_p(\omega)} + \|g\|_{L_p(\omega)} \\ &= \|f\|_{L_p(\omega)} + \|g\|_{L_p(\omega)}. \end{aligned}$$

So $f + g \in L_p(\omega)$. Now, note that

$$\begin{aligned} \|(f + g)^{p-1}\|_{L_q(\omega)} &= \left(\int_0^{+\infty} |f(x) + g(x)|^{q(p-1)}\omega(x)\, dx\right)^{\frac{1}{q}} \\ &= \left(\int_0^{+\infty} |f(x) + g(x)|^{p}\omega(x)\, dx\right)^{\frac{1}{q}} \\ &= \|f + g\|_{L_p(\omega)}^{p/q} < +\infty, \end{aligned}$$

therefore $(f + g)^{p-1} \in L_q(\omega)$. By Hölder's inequality, we have

$$\begin{aligned} &\int_0^{+\infty} |f(x) + g(x)|^{p}\omega(x)\, dx \\ &\quad = \int_0^{+\infty} |f(x) + g(x)|\,|f(x) + g(x)|^{p-1}\omega(x)\, dx \end{aligned}$$

$$\leq \int_0^{+\infty} |f(x)|\,|f(x)+g(x)|^{p-1}\omega(x)\,dx + \int_0^{+\infty} |g(x)|\,|f(x)+g(x)|^{p-1}\omega(x)\,dx$$

$$\leq \left[\left(\int_0^{+\infty} |f(x)|^p\omega(x)\,dx\right)^{\frac{1}{p}} + \left(\int_0^{+\infty} |g(x)|^q\omega(x)\,dx\right)^{\frac{1}{q}}\right]\left(\int_0^{+\infty} |f(x)+g(x)|^p\omega(x)\,dx\right)^{\frac{1}{q}}.$$

Then,

$$\left(\int_0^{+\infty} |f(x)+g(x)|^p\omega(x)\,dx\right)^{\frac{1}{p}} \leq \left(\int_0^{+\infty} |f(x)|^p\omega(x)\,dx\right)^{\frac{1}{p}} + \left(\int_0^{+\infty} |g(x)|^p\omega(x)\,dx\right)^{\frac{1}{p}},$$

and so

$$\|f+g\|_{L_p(\omega)} \leq \|f\|_{L_p(\omega)} + \|g\|_{L_q(\omega)}. \qquad \square$$

As an immediate consequence of the previous result, one has the following:

Corollary 3.1. $\|\cdot\|_{L_p(\omega)} : L_p(\omega) \to [0,+\infty]$ *is a norm.*

Corollary 3.2. $(L_p(\omega), \|\cdot\|_{L_p(\omega)})$ *is a normed space.*

The spaces $L_p(\omega)$ and $L_q(\omega)$ are, in general, not comparable, i. e., for $p \neq q$ there exist functions f, g such that $f \in L_p(\omega)$ and $f \notin L_q(\omega)$, and also $g \in L_q(\omega)$ but $g \notin L_p(\omega)$ (we invite the reader to provide such examples). However, if one restricts from $(0,+\infty)$ to a Lebesgue measurable set E of finite measure, then we can establish an embedding result. Accordingly, we define

$$L_p(E,\omega) := \{f\mathcal{X}_E \mid f \in L_p(\omega)\},$$

where $\mathcal{X}_E$ denotes the characteristic function on the set E, i. e.

$$\mathcal{X}_E(x) = \begin{cases} 1, & \text{if } x \in E, \\ 0, & \text{if } x \notin E. \end{cases}$$

Proposition 3.1. *Let* $1 \leq p \leq q \leq +\infty$*. Then* $L_q(E,\omega) \subset L_p(E,\omega)$ *for any measurable set* $E \subset (0,+\infty)$ *of finite measure.*

Proof. Let $f \in L_q(\omega)$ and $E \subset (0,+\infty)$ be a measurable set. Let $r = \frac{q}{p}$ and $s = \frac{r}{r-1}$. Then $\frac{1}{r} + \frac{1}{s} = 1$. Observe that

$$\int_0^{+\infty} |f(x)|^{pr}\,\omega(x)\,dx = \int_0^{+\infty} |f(x)|^q\,\omega(x)\,dx < +\infty.$$

By Hölder's inequality, we have

$$\int_0^{+\infty} |f(x)|^p \mathcal{X}_E(x)\,\omega(x)\,dx = \int_0^{+\infty} |f(x)|^p \mathcal{X}_E(x)[\omega(x)]^{\frac{1}{r}}[\omega(x)]^{\frac{1}{s}}\,\omega\,dx$$
$$\leq \left(\int_0^{+\infty} |f(x)|^q\,\omega(x)\,dx\right)^{\frac{1}{r}} \left(\int_E \omega(x)\,dx\right)^{\frac{1}{s}}$$
$$= (\omega(E))^{\frac{1}{s}} \left(\int_0^{+\infty} |f(x)|^q\,\omega(x)\,dx\right)^{\frac{1}{r}}. \qquad \square$$

The next fact is an interpolation result for weighted Lebesgue spaces.

Proposition 3.2. *If $1 < p < q < r$. Then $L_p(\omega) \cap L_r(\omega) \subset L_q(\omega)$ and*

$$\|f\|_{L_q(\omega)} \leq \|f\|^{\lambda}_{L_p(\omega)} \|f\|^{1-\lambda}_{L_r(\omega)},$$

where $0 < \lambda < 1$ and $\frac{1}{q} = \frac{\lambda}{p} + \frac{1-\lambda}{r}$.

Proof. Since $\frac{1}{q} = \frac{\lambda}{p} + \frac{1-\lambda}{r}$ for $0 < \lambda < 1$, we have $\frac{\lambda}{q} + \frac{(1-\lambda)q}{p} = 1$ and so, using Hölder's inequality, we get for $f \in L_p(\omega) \cap L_r(\omega)$,

$$\int_0^{+\infty} |f(x)|^q\,\omega(x)\,dx$$
$$= \int_0^{+\infty} |f(x)|^{\lambda q} |f(x)|^{(1-\lambda)q}\,\omega(x)\,dx$$
$$\leq \left(\int_0^{+\infty} |f(x)|^{(\lambda q)\frac{p}{\lambda q}}\,\omega(x)\,dx\right)^{\frac{\lambda q}{p}} \left(\int_0^{+\infty} (|f(x)|^{(1-\lambda)q})^{\frac{r}{(1-\lambda)q}}\,\omega(x)\,dx\right)^{\frac{(1-\lambda)q}{r}}.$$

Hence

$$\left(\int_0^{+\infty} |f(x)|^q\,\omega(x)\,dx\right)^{\frac{1}{q}} \leq \left[\left(\int_0^{+\infty} |f(x)|^p\,\omega(x)\,dx\right]^{\frac{1}{p}}\right)^{\lambda} \left[\left(\int_0^{+\infty} |f(x)|^r\,\omega(x)\,dx\right)^{\frac{1}{r}}\right]^{1-\lambda}.$$

Therefore,

$$\|f\|_{L_q(\omega)} \leq \|f\|^{\lambda}_{L_p(\omega)} \|f\|^{1-\lambda}_{L_q(\omega)}$$

and $f \in L_q(\omega)$, thus $L_p(\omega) \cap L_r(\omega) \subset L_q(\omega)$. $\square$

The $L_p(\omega)$ norm of a function f may be obtained by means of an infimum. This method is known as the *Luxemburg norm* of f. The precise result is stated as follows.

Theorem 3.3 (Luxemburg norm). *Let $f \in L_p(\omega)$ where $1 \le p < +\infty$. Then*

$$\|f\|_{L_p(\omega)} = \inf\left\{\lambda > 0 : \int_0^{+\infty} \left|\frac{f(x)}{\lambda}\right|^p \omega(x)\,dx \le 1\right\}. \tag{3.3}$$

Proof. On the one hand, let us take $\lambda = \|f\|_{L_p(\omega)}$, then

$$\inf\left\{\lambda > 0 : \int_0^{+\infty} \left|\frac{f(x)}{\lambda}\right|^p \omega(x)\,dx \le 1\right\} \le \|f\|_{L_p(\omega)}. \tag{3.4}$$

On the other hand, if $\int_0^{+\infty} |\frac{f(x)}{\lambda}|^p \omega(x)\,dx \le 1$, then $\int_0^{+\infty} |f(x)|^p \omega(x)\,dx \le \lambda^p$ and thus $(\int_0^{+\infty} |f(x)|^p \omega(x)\,dx)^{\frac{1}{p}} \le \lambda$. Therefore,

$$\|f\|_{L_p(\omega)} \le \inf\left\{\lambda > 0 : \int_0^{+\infty} \left|\frac{f(x)}{\lambda}\right|^p \omega(x)\,dx \le 1\right\}. \tag{3.5}$$

Combining (3.4) and (3.5), we have (3.3). □

Theorem 3.4. *$(L_p(\omega), \|\cdot\|_{L_p(\omega)})$ is a Banach space.*

Proof. Let $\{f_n\}_{n\in\mathbb{N}}$ be a Cauchy sequence in $L_p(\omega)$. Let us choose $\epsilon_1(\frac{\lambda_0}{\epsilon})^p < \frac{1}{n+m}$ for $n, m \in \mathbb{N}$ and $\epsilon > 0$, $\lambda_0 > 0$. For such ϵ_1, there exists an $n_0 \in \mathbb{N}$ such that

$$\|f_n - f_m\|_{L_p(\omega)} < \epsilon_1$$

whenever $n, m \ge n_0$. Considering the norm defined by (3.3), we can use $\lambda_0 > 0$ in such a way that $\lambda_0 < \epsilon_1$ and

$$\int_0^{\infty} \frac{|f_n(x) - f_m(x)|^p}{\lambda_0} \omega(x)\,dx \le 1.$$

Let $E = \{x \in X : |f_n(x) - f_m(x)| > \epsilon\}$, then

$$\epsilon\chi_E(x) \le |f_n(x) - f_m(x)|.$$

Therefore,

$$\int_0^{\infty} \left(\frac{\epsilon\chi_E}{\lambda_0}\right)^p \omega(x)\,dx \le \int_0^{\infty} \left|\frac{f_n(x) - f_m(x)}{\lambda_0}\right|^p \omega(x)\,dx.$$

Then

$$\left(\frac{\epsilon}{\lambda_0}\right)^p \int_0^{\infty} \chi_E(x)\omega(x)\,dx \le \int_0^{\infty} \left|\frac{f_n(x) - f_m(x)}{\lambda_0}\right|^p \omega(x)\,dx.$$

Thus

$$\left(\frac{\epsilon}{\lambda_0}\right)^p \int_E \omega(x)\,dx \le \int_0^\infty \left|\frac{f_n(x)-f_m(x)}{\lambda_0}\right|^p \omega(x)\,dx.$$

So

$$\left(\frac{\epsilon}{\lambda_0}\right)^p \omega(E) \le \int_0^\infty \left|\frac{f_n(x)-f_m(x)}{\lambda_0}\right|^p \omega(x)\,dx.$$

Then

$$\begin{aligned}\omega(E) &\le \left(\frac{\lambda_0}{\epsilon}\right)^p \int_0^\infty \left|\frac{f_n(x)-f_m(x)}{\lambda_0}\right|^p \omega(x)\,dx \\ &\le \left(\frac{\lambda_0}{\epsilon}\right)^p \epsilon_1 < \frac{1}{n+m}\end{aligned}$$

and so

$$\lim_{n,m\to\infty} \omega(\{x \in X : |f_n(x)-f_m(x)| > \epsilon\}) \le \lim_{n,m\to\infty} \frac{1}{n+m} = 0.$$

Thus

$$\lim_{n,m\to\infty} \omega(\{x \in X : |f_n(x)-f_m(x)| > \epsilon\}) = 0,$$

which means that $\{f_n\}_{n\in\mathbb{N}}$ is a Cauchy sequence in measure. Then some subsequence of $\{f_n\}_{n\in\mathbb{N}}$, say $\{f_{n_k}\}_{k\in\mathbb{N}}$, converges almost everywhere to a measurable function f, that is, $f_{n_k} \to f$ μ-a. e. as $k \to \infty$. Finally, by Fatou's lemma, one has

$$\int_0^\infty \left|\frac{f_n - f_m}{\lambda_0}\right|^p \omega(x)\,dx \le \liminf_{n\to\infty} \int_0^\infty \left|\frac{f_n(x)-f_m(x)}{\lambda_0}\right|^p \omega(x)\,dx \le 1$$

whenever $n \ge n_0$. Therefore $f_n - f_m$ belongs to $L_p(\omega)$. Since $L_p(\omega)$ is a linear space, one has

$$f = (f - f_n) + f_n \in L_p(\omega).$$

Moreover,

$$\limsup_{n\to\infty} \int_0^\infty \left|\frac{f - f_m}{\lambda_0}\right|^p \omega(x)\,dx \le 1$$

whenever $n \ge n_0$. Altogether we have

$$\lim_{n\to\infty} \|f_n - f\|_{L_p(\omega)} = 0.$$

This proves that $L_p(\omega)$ is complete, and so a Banach space. □

3.1 Duality

Duality on the weighted Lebesgue space $L_p(\omega)$ with $1 < p < +\infty$ is defined by means of the inner product

$$\langle f, g\rangle := \int_0^{+\infty} f(x)\, g(x)\, dx, \quad f \in L_p(\omega).$$

Theorem 3.5. *Let $\frac{1}{p} + \frac{1}{q} = 1$ and $g \in L_q(\omega^{1-q})$. Then $\sup_{\|f\|_{L_p(\omega)}=1} |\langle f, g\rangle| = \|g\|_{L_q(\omega^{1-q})}$ and $(L_p(\omega))^* = L_q(\omega^{1-q})$.*

Proof. By Hölder's inequality we have

$$\begin{aligned}
|\langle f, g\rangle| &\le \int_0^{+\infty} |f(x)|\, (\omega(x))^{\frac{1}{p}}\, |g(x)|\, (\omega(x))^{-\frac{1}{p}}\, dx \\
&\le \left(\int_0^{+\infty} |f(x)|^p\, \omega(x)\, dx\right)^{\frac{1}{p}} \left(\int_0^{+\infty} |g(x)|^q\, (\omega(x))^{-\frac{q}{p}}\, dx\right)^{\frac{1}{p}} \\
&\le \|f\|_{L_p(\omega)}\, \|g\|_{L_q(\omega^{1-q})}.
\end{aligned}$$

And so

$$\sup_{\|f\|_{L_p(\omega)}=1} |\langle f, g\rangle| \le \|g\|_{L_q(\omega^{1-q})}. \tag{3.6}$$

Next, since $\frac{q}{p} = q - 1$ and if

$$f = \frac{|g|^{q-1} (\operatorname{sgn} g) \omega^{1-q}}{\|g\|_{L_q(\omega^{1-q})}^{q-1}},$$

then $\|f\|_{L_p(\omega)} = 1$ and $\langle f, g\rangle = \|g\|_{L_q(\omega^{1-q})}$. Hence

$$\|g\|_{L_q(\omega^{1-q})} \le \sup_{\|f\|_{L_p(\omega)}=1} |\langle f, g\rangle| \le \|g\|_{L_q(\omega^{1-q})},$$

and the result follows. □

The $L_p(\omega)$ norm of f may be obtained via duality, as the next result shows.

Theorem 3.6. *If $f \in L_p(\omega)$ and $1 \le p < \infty$. Then*

$$\|f\|_{L_p(\omega)} = \sup_{\|g\|_{L_q(\omega)}=1} \left| \int_0^{+\infty} f(x) g(x) \omega(x)\, dx \right|,$$

where $\frac{1}{p} + \frac{1}{q} = 1$.

Proof. Let $g \in L_q(\omega)$ with $\|g\|_{L_q(\omega)} = 1$. Then by Theorem 3.1 (Hölder's inequality),

$$\left|\int_0^{+\infty} f(x)g(x)\omega(x)\,dx\right| \le \|f\|_{L_p(\omega)}\|g\|_{L_q(\omega)}$$

so

$$\sup_{\|g\|_{L_q(\omega)}}\left|\int_0^{+\infty} f(x)g(x)\omega(x)\,dx\right| \le \|f\|_{L_p(\omega)}.$$

Now, letting

$$g(x) = \frac{|f(x)|^{p-1}\operatorname{sgn} f(x)}{\|f\|_{L_p(\omega)}^{p-1}},$$

we have

$$\begin{aligned}\int_0^{+\infty} |g(x)|^q\omega(x)\,dx &= \int_0^{+\infty}\left|\frac{|f(x)|^{p-1}\operatorname{sgn} f(x)}{\|f\|_{L_p(\omega)}^{p-1}}\right|^q\omega(x)\,dx\\ &= \frac{1}{\|f\|_{L_p(\omega)}^{q(p-1)}}\int_1^{+\infty}|f(x)|^{q(p-1)}\omega(x)\,dx\\ &= \frac{1}{\|f\|_{L_p(\omega)}^{p}}\int_0^{+\infty}|f(x)|^{p}\omega(x)\,dx = 1.\end{aligned}$$

Also

$$\begin{aligned}\left|\int_0^{\infty} f(x)g(x)\omega(x)\,dx\right| &= \left|\int_0^{+\infty} f(x)\frac{|f(x)|^{p-1}\operatorname{sgn} f(x)}{\|f\|_{L_p(\omega)}^{p-1}}\omega(x)\,dx\right|\\ &= \left|\int_0^{+\infty}\frac{f(x)\operatorname{sgn} f(x)|f(x)|^{p-1}\omega(x)\,dx}{\|f\|_{L_p(\omega)}^{p-1}}\right|\\ &= \frac{1}{\|f\|_{L_p(\omega)}^{p-1}}\int_0^{+\infty}|f(x)|^{p}\omega(x)\,dx\\ &= \|f\|_{L_p(\omega)}.\end{aligned}$$

Then we have

$$\begin{aligned}\|f\|_{L_p(\omega)} &= \left|\int_0^{+\infty} f(x)g(x)\omega(x)\,dx\right|\\ &\le \sup_{\|g\|_{L_q(\omega)}=1}\left|\int_0^{+\infty} f(x)g(x)\omega(x)\,dx\right| \le \|f\|_{L_p(\omega)},\end{aligned}$$

and so

$$\|f\|_{L_p(\omega)} = \sup_{\|g\|_{L_q(\omega)}=1} \left| \int_0^{+\infty} f(x)g(x)\omega(x)\,dx \right|. \qquad \square$$

3.2 Cavalieri's principle in $L_p(\omega)$

We now obtain a Cavalieri's-type result for the weighted Lebesgue decreasing functions. Recall that m stands for the Lebesgue measure on $\mathbb{R}^n$.

Theorem 3.7. *If $f \in (L_p(\omega))^d$ then, for any measurable weight function ω, we have*

$$\int_0^{+\infty} |f(x)|^p \omega(x)\,dx = p \int_0^{+\infty} \lambda^{p-1} \left(\int_0^{D_f(\lambda)} \omega(x)\,dx \right) d\lambda,$$

where $D_f(\lambda) := m(\{x : |f(x)| > \lambda\})$.

Proof. First for all, notice that since $f \in (L_p(\omega))^d$, then

$$\{t \in (0,+\infty) : f(t) > \lambda\} = (0, D_f(\lambda)).$$

Let us define $E_f(\lambda) := \{t \in (0,+\infty) : f(t) > \lambda\}$.

Then $D_f(\lambda) = m(E_f(\lambda))$. Now, by Fubini's theorem, we obtain

$$\begin{aligned}
\int_0^{+\infty} |f(x)|^p \omega(x)\,dx &= \int_0^{+\infty} \left(p \int_0^{|f(x)|} \lambda^{p-1}\,d\lambda \right) \omega(x)\,dx \\
&= p \int_0^{+\infty} \left(\int_0^{+\infty} \lambda^{p-1} \chi_{E_f(\lambda)}(x)\,d\lambda \right) \omega(x)\,dx \\
&= p \int_0^{+\infty} \lambda^{p-1} \left(\int_0^{+\infty} \chi_{E_f(\lambda)}(x)\,\omega(x)\,dx \right) d\lambda \\
&= p \int_0^{+\infty} \lambda^{p-1} \left(\int_{E_f(\lambda)} \omega(x)\,dx \right) d\lambda \\
&= p \int_0^{+\infty} \lambda^{p-1} \left(\int_{(0,D_f(\lambda))} \omega(x)\,dx \right) d\lambda \\
&= p \int_0^{+\infty} \lambda^{p-1} \int_0^{D_f(\lambda)} \omega(x)\,dx\,d\lambda,
\end{aligned}$$

which ends the proof. $\square$

Theorem 3.7 is the weighted version of a result which is called the Cavalieri's principle, also known as the layer cake representation theorem (see Figure 3.1). It is quite

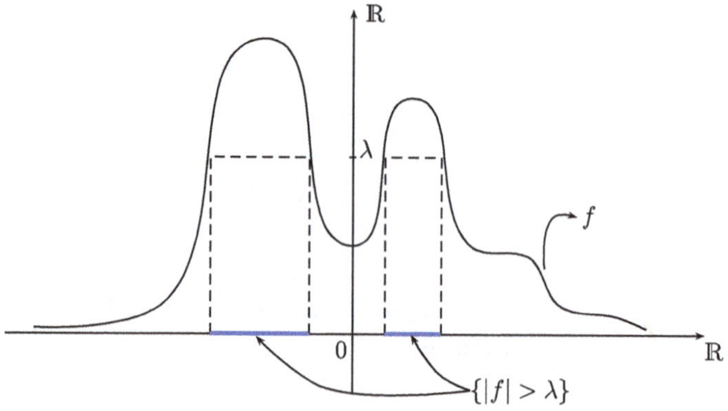

Figure 3.1: Cavalieri's principle.

important. For example, it allows us to calculate an integral of a multivariate function, via a one-dimensional integral. For the sake of completeness and the convenience of the reader, we state and prove this famous result.

Theorem 3.8. *Let f be a Lebesgue measurable function. Then*

$$\int_{\mathbb{R}^n} |f|\, dm = \int_0^{\infty} m\big(\{x \in \mathbb{R}^n : |f(x)| > \lambda\}\big)\, d\lambda. \tag{3.7}$$

Proof. If $m(\{x \in \mathbb{R}^n : |f(x)| > \lambda\}) = \infty$, there is nothing to prove. Therefore, let us suppose that $m(\{x \in \mathbb{R}^n : |f(x)| > \lambda\}) < \infty$. We want to show that $\{x \in \mathbb{R}^n : |f(x)| > \lambda\}$ is measurable over $[0, \infty)$. Let us consider the set

$$E = \{(x, \lambda) \in \mathbb{R}^n \times [0, \infty) : 0 < \lambda < |f(x)|\}.$$

We shall show that E is a measurable set in $\mathbb{R}^n \times [0, \infty)$.

Since $|f(x)| \geq 0$, there is a sequence $\{S_n\}_{n \in \mathbb{N}}$ of simple functions such that $0 \leq S_n \leq S_{n+1} \leq |f|$ and $\lim_{n \to \infty} S_n = |f|$. Hence, S_n can be written as

$$S_n = \sum_{j=1}^{\infty} a_j^n \chi_{A_j^n},$$

with each A_j^n disjoint and measurable, and $j = 1, \ldots, n$.

We shall prove that

$$\begin{aligned} E_n &= \{(x, \lambda) \in \mathbb{R}^n \times [0, \infty) : 0 < \lambda < S_n(x)\} \\ &= \bigcup_{j=1}^{\infty} A_j^n \times (0, a_j^n). \end{aligned}$$

Indeed, let $(x,\lambda) \in E_n$ then $0 < \lambda < S_n(x)$ and $x \in A_j^n$ for some j. Hence $S_n(x) = a_j^n$ and so $0 < \lambda < a_j^n$, thus $\lambda \in (0, a_j^n)$. And so

$$(x,\lambda) \in A_j^n \times (0, a_j^n)$$

for some j, yielding

$$(x,\lambda) \in \bigcup_{j=1}^{\infty} A_j^n \times (0, a_j^n),$$

which implies

$$E_n \subset \bigcup_{j=1}^{\infty} A_j^n \times (0, a_j^n). \tag{3.8}$$

Now, let $(x,\lambda) \in \bigcup_{j=1}^{\infty} A_j^n \times (0, a_j^n)$ for some j. Thus $x \in A_j^n$ and $0 < \lambda < a_j^n$. Hence $x \in A_j^n$ implies that $\chi_{A_j^n}(x) = 1$ and so

$$0 < \lambda < a_j^n \chi_{A_j^n}(x) \le \sum_{j=1}^{\infty} a_j^n \chi_{A_j^n}(x) = S_n(x),$$

which means that

$$(x,\lambda) \in E_n.$$

Thus

$$\bigcup_{j=1}^{\infty} A_j^n \times (0, a_j^n) \subset E_n. \tag{3.9}$$

In this way, we have shown that

$$\begin{aligned} E_n &= \{(x,\lambda) \in \mathbb{R}^n \times [0,\infty) : 0 < \lambda < S_n(x)\} \\ &= \bigcup_{j=1}^{\infty} A_j^n \times (0, a_j^n). \end{aligned}$$

From the latter equality, we conclude that E_n is measurable.

And now let $(x,\lambda) \in \mathbb{R}^n \times (0,\infty)$. If $|f(x)| = 0$, since

$$0 < S_n \le S_{n+1}(x) \le |f(x)| = 0,$$

one has $S_n(x) = 0$ for all $n \in \mathbb{N}$, which means that $(x,\lambda) \notin E_n$, and thus $\chi_{E_n}(x,\lambda) = 0$ for all $n \in \mathbb{N}$. Also, if $|f(x)| = 0$, then $(x,\lambda) \notin E$ and so $\chi_E(x,\lambda) = 0$. Therefore,

$$\lim_{n\to\infty} \chi_{E_n}(x,\lambda) = \chi_E(x,\lambda).$$

Thus if $|f(x)| > 0$, then there exists $\lambda > 0$ such that $0 < \lambda < |f(x)|$ and

$$\chi_E(x, \lambda) = 1.$$

Since $\lim_{n\to\infty} S_n(x) = |f(x)|$, there exist $N, \delta > 0$ such that

$$0 < \lambda \leq \delta \leq S_n \leq |f(x)|$$

for all $n \geq \mathbb{N}$, thus $\chi_{E_n}(x, \lambda) = 1$, and so

$$\lim_{n\to\infty} \chi_{E_n}(x, \lambda) = \chi_E(x, \lambda).$$

Since χ_E is the limit of a sequence of measurable functions, it is also a measurable function, hence E is measurable, too.

The measurability of f easily implies that the set

$$E^\lambda = \{x \in \mathbb{R}^n : |f(x)| > \lambda\}$$

is measurable on $[0, \infty)$. Next, applying Fubini's theorem, we obtain

$$\begin{aligned}
\int_{\mathbb{R}^n} |f|\, dm &= \int_{\mathbb{R}^n} \int_0^{|f|} d\lambda dm \\
&= \int_{\mathbb{R}^n} \int_0^{\infty} \chi_{[0,|f|]}(\lambda)\, d\lambda dm \\
&= \int_0^{\infty} \int_{\mathbb{R}^n} \chi_{\{x\in\mathbb{R}^n:|f(x)|>\lambda\}}\, dm d\lambda \\
&= \int_0^{\infty} m(\{x \in \mathbb{R}^n : |f(x)| > \lambda\})\, d\lambda,
\end{aligned}$$

ending the proof. □

3.3 Hardy operator on $L_p(\omega)$ spaces

According to D. E. Edmunds, "Hardy-type operators play a fundamental role throughout analysis, and this has motivated a considerable amount of research into their properties when they act between a wide variety of function spaces" [35].

In this section, we study the Hardy operator acting between $L_p(\omega)$ spaces. Let us recall the definition of the Hardy operator.

Definition 3.3 (Hardy operator). Let f be a positive and measurable function on $(0, +\infty)$. The Hardy operator is defined as

$$Hf(x) := \frac{1}{x} \int_0^x f(t)\, dt.$$

In the following result, we will see that, for $1 < p < +\infty$, Hf is a bounded operator from $L_p(\omega)$ into $L_p(\omega)$.

Theorem 3.9. *Let $f \in L_p(\omega)$ with $1 < p < +\infty$. If ω is a nonincreasing weight, then*

$$\|Hf\|_{L_p(\omega)} \le \frac{p}{p-1}\|f\|_{L_p(\omega)}. \tag{3.10}$$

Proof. Making a convenient change of variable twice and using the Minkowski integral inequality, we have

$$\begin{aligned}
\left(\int_0^{+\infty}\left[\frac{1}{x}\int_0^x f(t)\,dt\right]^p \omega(x)\,dx\right)^{\frac{1}{p}} &= \left(\int_0^{+\infty}\left[\frac{1}{x}\int_0^1 f(xu)\,xdu\right]^p \omega(x)\,dx\right)^{\frac{1}{p}}\\
&= \left(\int_0^{+\infty}\left[\int_0^1 f(xu)\,du\right]^p \omega(x)\,dx\right)^{\frac{1}{p}}\\
&\le \int_0^1\left(\int_0^{+\infty}(f(xu))^p\,\omega(x)\,dx\right)^{\frac{1}{p}} du\\
&= \int_0^1\left(\int_0^{+\infty}(f(v))^p\,\omega\left(\frac{v}{u}\right)\frac{dv}{u}\right)^{\frac{1}{p}} du.
\end{aligned}$$

Since $0 < u < 1$, we have $\frac{1}{u} > 1$ and $\omega(\frac{v}{u}) \le \omega(v)$, since ω is nonincreasing. Therefore,

$$\begin{aligned}
\int_0^1\left(\int_0^{+\infty}(f(v))^p\,\omega\left(\frac{v}{u}\right)\frac{dv}{u}\right)^{\frac{1}{p}} du &\le \int_0^1\left(\int_0^{+\infty}(f(v))^p\,\omega(v)\,dv\right)^{\frac{1}{p}}\frac{du}{u^{1/p}}\\
&= \left(\int_0^1 u^{-\frac{1}{p}}\,du\right)\left(\int_0^{+\infty}(f(v))^p\,\omega(v)\,dv\right)^{\frac{1}{p}}\\
&= \frac{p}{p-1}\left(\int_0^{+\infty}(f(v))^p\,\omega(v)\,dv\right)^{\frac{1}{p}}.
\end{aligned}$$

Hence it follows that

$$\left(\int_0^{+\infty}\left[\frac{1}{x}\int_0^x f(t)\,dt\right]^p \omega(x)\,dx\right)^{\frac{1}{p}} \le \frac{p}{p-1}\left(\int_0^{+\infty}(f(x))^p\,\omega(x)\,dx\right)^{\frac{1}{p}}, \tag{3.11}$$

as needed. □

Corollary 3.3. *Under the same conditions as in Theorem 3.9, for $r \ge 0$, the inequality*

$$r^p\int_0^{+\infty}\frac{\omega(x)}{x^p}\,dx \le B\int_0^r \omega(x)\,dx$$

holds.

Proof. Take $f = \chi_{[0,r]}$ in (3.11). □

Now, we would like to study the adjoint of the Hardy operator. In order to do this, let us first recall the definition of an adjoint operator.

Definition 3.4. Let $T : X \to Y$ be a linear and bounded operator. We say that the operator $T^* : Y^* \to X^*$ is the adjoint operator of T if it satisfies the duality identity, which means that for all $x \in X, y \in Y^*$, where X and Y are Banach spaces, we have

$$\langle Tx, y\rangle = \langle x, T^* y\rangle,$$

where $\Lambda(\xi) = \langle \xi, \Lambda\rangle$ with $\Lambda \in \Sigma^*$ and $\xi \in \Sigma$.

We obtain now the adjoint of the Hardy operator.

Theorem 3.10. *The adjoint of the Hardy operator $H : L_p \to L_p$, at least formally, is given by*

$$H^* f(y) = \int_y^{+\infty} f(x) \frac{dx}{x} \quad \text{for } f \geq 0.$$

Proof. From Definition 3.4, Riesz representation theorem, and Fubini's theorem, we get

$$\begin{aligned}
\langle Hf, g\rangle = \int_0^{+\infty} Hf(x)\, g(x)\, dx &= \int_0^{+\infty} \left(\frac{1}{x} \int_0^x f(y)\, dy \right) g(x)\, dx \\
&= \int_0^{+\infty} \left(\frac{1}{x} \int_0^{+\infty} \mathcal{X}_{(0,x)}(y) f(y)\, dy \right) g(x)\, dx \\
&= \int_0^{+\infty} \left(\int_0^{+\infty} \mathcal{X}_{(y,+\infty)}(x) f(y)\, dy \right) \frac{g(x)}{x}\, dx \\
&= \int_0^{+\infty} f(y) \int_y^{+\infty} g(x) \frac{dx}{x}\, dy \\
&= \langle f, H^* g\rangle,
\end{aligned}$$

which ends the proof. □

The adjoint of the Hardy operator, namely H^*, is a bounded operator from $L_q(\omega)$ into $L_q(\omega)$ for $1 < q < +\infty$. More precisely, we have the following result.

Theorem 3.11. *Let $f \in L_p(\omega)$ with $1 < q < +\infty$. If ω is a nondecreasing weight, then*

$$\left\| H^* f \right\|_{L_q(\omega)} \leq \frac{p}{p-1} \|f\|_{L_q(\omega)}.$$

Proof. Observe that if $x = zy$, then $dx = y\,dz$, therefore

$$H^*f(y) = \int_y^{+\infty} f(x)\,\frac{dx}{x} = \int_1^{+\infty} \frac{f(zy)}{zy} y\,dz = \int_1^{+\infty} \frac{f(zy)}{z}\,dz.$$

Now using the integral Minkowski inequality, we obtain

$$\begin{aligned}
\left(\int_y^{+\infty} (H^*f(y))^q\,\omega(y)\,dy\right)^{\frac{1}{q}} &= \left(\int_0^{+\infty}\left(\int_y^{+\infty} f(x)\,\frac{dx}{x}\right)^q \omega(y)\,dy\right)^{\frac{1}{q}} \\
&= \left(\int_0^{+\infty}\left(\int_1^{+\infty} \frac{f(zy)}{z}\,dz\right)^q \omega(y)\,dy\right)^{\frac{1}{q}} \\
&\le \int_1^{+\infty}\left(\int_0^{+\infty}\left(\frac{f(zy)}{z}\right)^q \omega(y)\,dy\right)^{\frac{1}{q}} dz \\
&= \int_1^{+\infty}\left(\int_0^{+\infty}\left(\frac{f(u)}{z}\right)^q \omega\left(\frac{u}{z}\right)\frac{du}{z}\right)^{\frac{1}{q}} dz.
\end{aligned}$$

As $z > 1$, we have $\frac{1}{z} < 1$ and $\omega(\frac{u}{z}) \le \omega(u)$ since ω is nondecreasing. Therefore,

$$\begin{aligned}
\int_1^{+\infty}\left(\int_0^{+\infty}\left(\frac{f(u)}{z}\right)^q \omega\left(\frac{u}{z}\right)\frac{du}{z}\right)^{\frac{1}{q}} dz &\le \int_1^{+\infty}\left(\int_0^{+\infty}\left(\frac{f(u)}{z}\right)^q \omega(u)\,\frac{du}{z}\right)^{\frac{1}{q}} dz \\
&= \int_1^{+\infty}\left(\int_0^{+\infty} [(f(u))^q \omega(u)\,du]\,\frac{1}{z^{1+q}}\right)^{\frac{1}{q}} dz \\
&= \int_1^{+\infty} \frac{1}{z^{1+1/q}}\left(\int_0^{+\infty} [(f(u))^q \omega(u)\,du]\right)^{\frac{1}{q}} dz \\
&= \frac{p}{p-1}\left(\int_0^{+\infty} [(f(u))^q \omega(u)\,du]\right)^{\frac{1}{q}}.
\end{aligned}$$

Hence

$$\left(\int_0^{+\infty}\left(\int_y^{+\infty} f(x)\,\frac{dx}{x}\right)^q \omega(y)\,dy\right)^{\frac{1}{q}} \le \frac{p}{p-1}\left(\int_0^{+\infty} [(f(u))^q \omega(u)\,du]\right)^{\frac{1}{q}},$$

as required. □

3.4 Strict and uniform convexity of the $L_p(\omega)$ space

In this section we deal with some geometric properties of weighted Lebesgue space $L_p(\omega)$. We will basically follow the exposition in [60].

Let us first state, without proof, two lemmas which will be helpful in proving the convexity of $L_p(\omega)$.

Lemma 3.1. *Let $0 < p < 1$, then $(a+b)^p \le a^p + b^p$ for all $a \ge 0$ and $b \ge 0$.*

Lemma 3.2. *If $p \ge 1$, then $(a+b)^p \le 2^{p-1}(a^p + b^p)$ for all $a \ge 0$ and $b \ge 0$.*

Theorem 3.12. *The space $L_p(\omega)$ is convex whenever $0 < p < +\infty$.*

Proof. Let $f, g \in L_p(\omega)$. We are going to show that $\lambda f + (1-\lambda)g \in L_p(\omega)$ for $1 \le \lambda \le 1$. Let us consider this in two cases, $p \ge 1$ and $0 < p < 1$.

Case $p \ge 1$. By Lemma 3.2 and Minkowski inequality, we have

$$\begin{aligned}
&\int_0^{+\infty} |\lambda f(x) + (1-\lambda)g(x)|^p \omega(x)\,dx \\
&= \int_0^{+\infty} |(\lambda f(x) + (1-\lambda)g(x))\,(\omega(x))^{\frac{1}{p}}|^p\,dx \\
&= \left(\left(\int_0^{+\infty} |(\lambda f(x) + (1-\lambda)g(x))\,(\omega(x))^{\frac{1}{p}}|^p\,dx\right)^{\frac{1}{p}}\right)^p \\
&\le \left(\left(\int_0^{+\infty} |\lambda f(x)\,(\omega(x))^{\frac{1}{p}}|^p\,dx\right)^{\frac{1}{p}} + \left(\int_0^{+\infty} |(1-\lambda)g(x)\,(\omega(x))^{\frac{1}{p}}|^p\,dx\right)^{\frac{1}{p}}\right)^p \\
&\le 2^{p-1}\left(\int_0^{+\infty} |\lambda f(x)\,(\omega(x))^{\frac{1}{p}}|^p\,dx + \int_0^{+\infty} |(1-\lambda)g(x)\,(\omega(x))^{\frac{1}{p}}|^p\,dx\right) \\
&= 2^{p-1}\left(|\lambda|^p \int_0^{+\infty} |f(x)\,(\omega(x))^{\frac{1}{p}}|^p\,dx + |1-\lambda|^p \int_0^{+\infty} |g(x)\,(\omega(x))^{\frac{1}{p}}|^p\,dx\right) \\
&= 2^{p-1}(|\lambda|^p\,\|f\|^p_{L_p(\omega)} + |1-\lambda|^p\,\|g\|^p_{L_p(\omega)}) \\
&< +\infty,
\end{aligned}$$

which shows that $\lambda f + (1-\lambda)g \in L_p(\omega)$ for $1 \le p$.

Case $0 < p < 1$. Let $f, g \in L_p(\omega)$ and $\lambda \in [0,1]$. By Lemma 3.1, we have

$$\begin{aligned}
&\int_0^{+\infty} |\lambda f(x) + (1-\lambda)g(x)|^p \omega(x)\,dx \\
&= \int_0^{+\infty} |(\lambda f(x) + (1-\lambda)g(x))\,(\omega(x))^{\frac{1}{p}}|^p\,dx \\
&\leq \int_0^{+\infty} |\lambda f(x)\,(\omega(x))^{\frac{1}{p}}|^p\,dx + \int_0^{+\infty} |(1-\lambda)g(x)\,(\omega(x))^{\frac{1}{p}}|^p\,dx \\
&= |\lambda|^p \,\|f\|^p_{L_p(\omega)} + |1-\lambda|^p \,\|g\|^p_{L_p(\omega)} \\
&< +\infty,
\end{aligned}$$

completing the proof. □

Theorem 3.13. *The space $L_p(\omega)$ is strictly convex for $0 < p < +\infty$.*

Proof. Let $f, g \in L_p(\omega)$ with $f \neq g$, $\|f\|_{L_p(\omega)} = \|g\|_{L_p(\omega)} = 1$, and $0 \leq \lambda \leq 1$. The strict convexity of $L_p([0,+\infty))$ implies

$$\begin{aligned}
\|\lambda f + (1-\lambda)g\|_{L_p(\omega)} &= \left(\int_0^{+\infty} |(\lambda f(x) + (1-\lambda)g(x))\,(\omega(x))^{\frac{1}{p}}|^p\,dx\right)^p \\
&= \|(\lambda f + (1-\lambda)g)\omega^{\frac{1}{p}}\|_{L_p} \\
&< 1.
\end{aligned}$$

□

In order to prove the uniform convexity of $L_p(\omega)$, we will need the following lemmas.

Lemma 3.3. *Let $2 \leq p < +\infty$ and $a, b \in \mathbb{R}$. Then*

$$|a+b|^p + |a-b|^p \leq 2^{p-1}\left(|a|^p + |b|^p\right).$$

Lemma 3.4. *Let $2 \leq p < +\infty$ and $a, b \in \mathbb{R}$. For any $f, g \in L_p$, we have*

$$\|f+g\|^p_{L_p} + \|f-g\|^p_{L_p} \leq 2^{p-1}\left(\|f\|^p_{L_p} + \|g\|^p_{L_p}\right).$$

For a proof of Lemma 3.3, check [3]; and for a proof of Lemma 3.4, see [26]. The weighted version of the latter holds, as shown below.

Lemma 3.5. *Let $2 \leq p < +\infty$ and $a, b \in \mathbb{R}$. For any $f, g \in L_p(\omega)$, we have*

$$\|f+g\|^p_{L_p(\omega)} + \|f-g\|^p_{L_p(\omega)} \leq 2^{p-1}\left(\|f\|^p_{L_p(\omega)} + \|g\|^p_{L_p(\omega)}\right).$$

Proof. Let $f, g \in L_p(\omega)$. Then $f\omega^{1/p}, g\omega^{1/p} \in L_p$ and, by Lemma 3.5, we get

$$\begin{aligned}\|f+g\|^p_{L_p(\omega)} + \|f-g\|^p_{L_p(\omega)} &= \|f\omega^{1/p} + g\omega^{1/p}\|^p_{L_p} + \|f\omega^{1/p} - g\omega^{1/p}\|^p_{L_p} \\ &\le 2^{p-1}(\|f\omega^{1/p}\|^p_{L_p} + \|g\omega^{1/p}\|^p_{L_p}) \\ &= 2^{p-1}(\|f\|^p_{L_p(\omega)} + \|g\|^p_{L_p(\omega)}).\end{aligned}$$

□

We are ready to prove that the $L_p(\omega)$ space is uniformly convex. Let us begin with the case $2 \le p < +\infty$.

Theorem 3.14. *The space $L_p(\omega)$ is uniformly convex for $2 \le p < +\infty$.*

Proof. Let $f, g \in L_p(\omega)$ with $f \ne g$, $\|f\|_{L_p(\omega)} \le 1$, $\|g\|_{L_p(\omega)} \le 1$, and $\|f-g\|_{L_p(\omega)} \ge \epsilon$.

Then we have

$$\|f+g\|^p_{L_p(\omega)} \le 2^{p-1}(\|f\|^p_{L_p(\omega)} + \|g\|^p_{L_p(\omega)}) - \|f-g\|^p_{L_p(\omega)},$$

which implies that

$$\begin{aligned}\|f+g\|^p_{L_p(\omega)} &\le 2^{p-1}2 - \epsilon^p \\ &= 2^p\left(1-\left(\frac{\epsilon}{2}\right)^p\right).\end{aligned}$$

Therefore, we obtain

$$\begin{aligned}\left\|\frac{f+g}{2}\right\|^p_{L_p(\omega)} &\le 1-\left(\frac{\epsilon}{2}\right)^p \\ &\le 1-\delta(\epsilon),\end{aligned}$$

where

$$\delta(\epsilon) = 1-\left(1-\left(\frac{\epsilon}{2}\right)^p\right)^{1/p}.$$

□

Now, we would like to prove that the $L_p(\omega)$ space is uniformly convex for $1 < p < 2$. We gather some preliminary results.

Lemma 3.6 (Minkowski inequality for $p \in (0,1)$). *Let $0 < p < 1$ and f and g be positive functions in L_p. Then $f+g \in L_p$ and*

$$\|f+g\|_{L_p} \ge \|f\|_{L_p} + \|g\|_{L_p}.$$

For a proof of the above lemma, see [16, Corollary 3.82]. For a proof the lemma below, check [26].

Lemma 3.7. *If $1 < p < 2$ and $q = \frac{p}{p-1}$, then*

$$|a+b|^q + |a-b|^q \le 2\left(|a|^p + |b|^p\right)^{q-1}.$$

The following result easily follows from Lemma 3.7.

Lemma 3.8. *Let $1 < p < 2$ and $q = \frac{p}{p-1}$. For any $f, g \in L_p$, we have*

$$\|f+g\|_{L_p}^q + \|f-g\|_{L_p}^q \le 2\left(\|f\|_{L_p}^p + \|g\|_{L_p}^p\right)^{q-1}.$$

We state and prove the weighted version of Lemma 3.8.

Theorem 3.15. *Let $1 < p < 2$ and $q = \frac{p}{p-1}$. For any $f, g \in L_p(\omega)$, we have*

$$\|f+g\|_{L_p(\omega)}^q + \|f-g\|_{L_p(\omega)}^q \le 2\left(\|f\|_{L_p(\omega)}^p + \|g\|_{L_p(\omega)}^p\right)^{q-1}.$$

Proof. First note that

$$\begin{aligned}
\|f\|_{L_p(\omega)}^q &= \left(\left(\int_0^{+\infty} |f(x)|^p \omega(x)\,dx\right)^{\frac{1}{p}}\right)^q \\
&= \left(\int_0^{+\infty} |f(x)|^p \omega(x)\,dx\right)^{\frac{1}{p-1}} \\
&= \left(\int_0^{+\infty} |f(x)|^{q(p-1)} \omega(x)\,dx\right)^{\frac{1}{p-1}} \\
&= \|\,|f|^q\,\|_{L_{p-1}(\omega)}.
\end{aligned}$$

Next, let $F, G \in L_r$. By Minkowski inequality, for $0 < r < 1$ we have

$$\left(\int_0^{+\infty} |F(x)|^r\,dx\right)^{\frac{1}{r}} + \left(\int_0^{+\infty} |G(x)|^r\,dx\right)^{\frac{1}{r}} \le \left(\int_0^{+\infty} |F(x)+G(x)|^r\,dx\right)^{\frac{1}{r}}.$$

Let $f, g \in L_p(\omega)$. Since $1 < p < +\infty$, we have $0 < \frac{p}{q} < 1$. Let us define

$$F(x) = |(f(x)+g(x))(\omega(x))^{\frac{1}{p}}|^q$$

and

$$G(x) = |(f(x) - g(x))(\omega(x))^{\frac{1}{p}}|^q.$$

Then,

$$\begin{aligned}
&\left(\int_0^{+\infty} |(f(x) + g(x))(\omega(x))^{\frac{1}{p}}|^p \, dx\right)^{\frac{q}{p}} + \left(\int_0^{+\infty} |(f(x) - g(x))(\omega(x))^{\frac{1}{p}}|^p \, dx\right)^{\frac{q}{p}} \\
&\leq \left(\int_0^{+\infty} (|(f(x) + g(x))(\omega(x))^{\frac{1}{p}}|^q + |(f(x) - g(x))(\omega(x))^{\frac{1}{p}}|^q)^{\frac{p}{q}} \, dx\right)^{\frac{q}{p}} \\
&= \left(\int_0^{+\infty} (|f(x)(\omega(x))^{\frac{1}{p}} + g(x)(\omega(x))^{\frac{1}{p}}|^q + |f(x)(\omega(x))^{\frac{1}{p}} - g(x)(\omega(x))^{\frac{1}{p}}|^q)^{\frac{p}{q}} \, dx\right)^{\frac{q}{p}} \\
&\leq \left(\int_0^{+\infty} (2(|f(x)(\omega(x))^{\frac{1}{p}}|^p + |g(x)(\omega(x))^{\frac{1}{p}}|^p)^{q-1})^{\frac{p}{q}} \, dx\right)^{\frac{q}{p}} \\
&= 2\left(\int_0^{+\infty} (|f(x)(\omega(x))^{\frac{1}{p}}|^p + |g(x)(\omega(x))^{\frac{1}{p}}|^p) \, dx\right)^{\frac{q}{p}} \\
&= 2\left(\int_0^{+\infty} |f(x)|^p \omega(x) \, dx + \int_0^{+\infty} |g(x)|^p (\omega(x)) \, dx\right)^{\frac{q}{p}}.
\end{aligned}$$

Thus, we obtain

$$\|f + g\|_{L_p(\omega)}^q + \|f - g\|_{L_p(\omega)}^q \leq 2(\|f\|_{L_p(\omega)}^p + \|g\|_{L_p(\omega)}^p)^{q-1}. \qquad \square$$

Theorem 3.16. *The space $L_p(\omega)$ is uniformly convex for $1 < p < 2$.*

Proof. Let $f, g \in L_p(\omega)$, $1 < p < 2$, with $\|f\|_{L_p(\omega)} \leq 1$, $\|g\|_{L_p(\omega)} \leq 1$, and $\|f - g\|_{L_p(\omega)} \geq \epsilon$. Then by Theorem 3.15, we have

$$\begin{aligned}
\|f + g\|_{L_p(\omega)}^q &\leq 2(\|f\|_{L_p(\omega)}^p + \|g\|_{L_p(\omega)}^p)^{q-1} - \|f - g\|_{L_p(\omega)}^q \\
&\leq 2 \cdot 2^{q-1} - \epsilon^q \\
&= 2^q\left(1 - \left(\frac{\epsilon}{2}\right)^q\right).
\end{aligned}$$

Hence, we get

$$\left\| \frac{f+g}{2} \right\|_{L_p(\omega)}^q \leq \left(1 - \left(\frac{\epsilon}{2}\right)^q\right)^{\frac{1}{q}}$$
$$\leq 1 - \delta(\epsilon),$$

where $\delta(\epsilon) = 1 - (1 - (\frac{\epsilon}{2})^q)^{\frac{1}{q}}$. □

3.5 Calderón–Zygmund decomposition

A fundamental tool in harmonic analysis and singular integral theory, namely the Calderón–Zygmund decomposition, provides a powerful method for breaking down functions into structured components that reveal their fine-scale behavior. Originating in the pioneering work of Alberto Calderón and Antoni Zygmund, this technique has become indispensable in the study of Lebesgue spaces, maximal functions, and operators of singular integral type.

At its core, the Calderón–Zygmund decomposition splits a given function into a "good" part with controlled smoothness and a "bad" part supported on a carefully chosen family of cubes. This decomposition allows analysts to isolate singularities, control oscillations, and establish key estimates in the theory of L_p spaces. Its applications extend far beyond its initial context, playing a crucial role in the proofs of boundedness for singular integrals, the study of Hardy spaces, and even partial differential equations.

Its elegance and versatility make it an essential technique for both theoretical and applied mathematics. For further reading, see [40, 65].

3.5.1 Cubes

A closed cube is a bounded interval in $\mathbb{R}^n$ whose sides are parallel to the coordinate axes and equally long, that is,

$$Q = [a_1, b_1] \times [a_2, b_2] \times \cdots \times [a_n, b_n],$$

with $b_1 - a_1 = b_2 - a_2 = \cdots = b_n - a_n$. The side length of a cube Q is denoted by $l(Q)$. In case we want to specify the center, we write

$$Q(x, l) := \left\{ y \in \mathbb{R}^n : |y_k - x_k| \leq \frac{l}{2}, k = 1, 2, \ldots, n \right\},$$

for a cube with center at $x \in \mathbb{R}^n$ and side length $l > 0$. If $Q = Q(x, l)$, we denote $aQ = Q(x, al)$ for $a > 0$.

Thus aQ is the cube with the same center as Q, but the side length multiplied by the factor a. The integral average of $f \in L^1_{\text{loc}}(\mathbb{R}^n)$ in a cube Q is denoted by

$$f_Q = \fint_Q f(x)\,dx = \frac{1}{m(Q)} \int_Q f(x)\,dx.$$

Let $Q = [a_1, b_1] \times [a_2, b_2] \times \cdots \times [a_n, b_n]$ be a closed cube in $\mathbb{R}^n$ with side length l. We decompose Q into subcubes recursively. Denote $\mathcal{D}_0 = \{Q\}$. Bisect each interval $[a_i, b_i]$, $i = 1, 2, \ldots$, and obtain 2^n congruent subcubes of Q. Denote this collection of cubes by $\mathcal{D}_1$. Bisect every cube in $\mathcal{D}_1$ and obtain 2^n subcubes. By continuing this way, we obtain generations of dyadic cubes $\mathcal{D}_k$, $k = 0, 1, 2, \ldots$.

The dyadic subcubes in $\mathcal{D}_k$ are of the form

$$\left[a_1 + \frac{m_1 l}{2^k}, b_1 + \frac{(m_1+1)l}{2^k}\right] \times \left[a_2 + \frac{m_2 l}{2^k}, b_2 + \frac{(m_2+1)l}{2^k}\right] \times \cdots \times \left[a_n + \frac{m_1 l}{2^k}, b_n + \frac{(m_n+1)l}{2^k}\right],$$

where $k = 0, 1, 2, \ldots$, and $m_j = 0, 1, \ldots, 2^k - 1$, $j = 1, 2, \ldots, n$ (see Figure 3.2).

The collection of all dyadic subcubes of Q is

$$\mathcal{D} = \bigcup_{k=0}^{+\infty} \mathcal{D}_k. \tag{3.12}$$

A cube $Q' \in \mathcal{D}$ is called a dyadic subcube of Q.

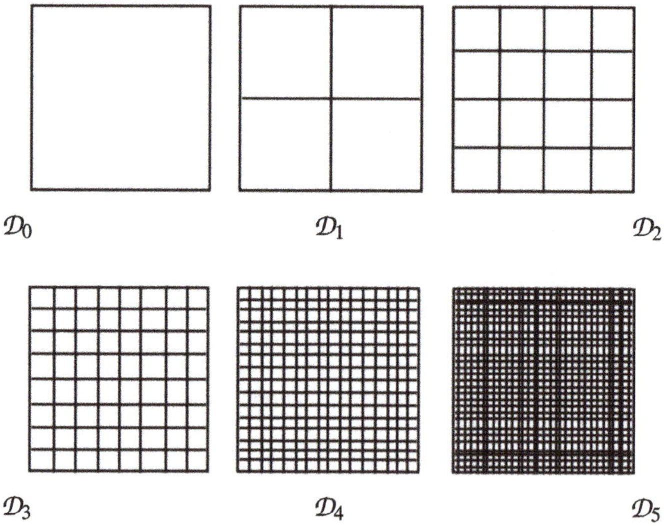

Figure 3.2: Collections of dyadic subcubes.

3.5.2 Properties of dyadic subcubes

Dyadic subcubes of Q have the following properties:

1. Every $Q' \in \mathcal{D}$ is a subcube of Q.
2. Cubes in $\mathcal{D}_k$ cover Q and the interiors of the cubes in $\mathcal{D}_k$ are pairwise disjoint for every $k = 0, 1, 2, \ldots$.
3. If $Q', Q'' \in \mathcal{D}$, either one is contained in the other or the interiors of the cubes are disjoint. This is called the nestedness property, see Figure 3.3.
4. If $Q' \in \mathcal{D}$ and $j < k$, there is exactly one parent cube in $\mathcal{D}_j$ which contains Q'.
5. Every cube $Q' \in \mathcal{D}_k$ is a union of exactly 2^n children cubes $Q'' \in \mathcal{D}_{k+1}$ with $m(Q') = 2^m M(Q'')$.
6. If $Q' \in \mathcal{D}_k$, then $l(Q') = 2^{-k} l(Q)$ and $m(Q') = 2^{-nk} m(Q)$.

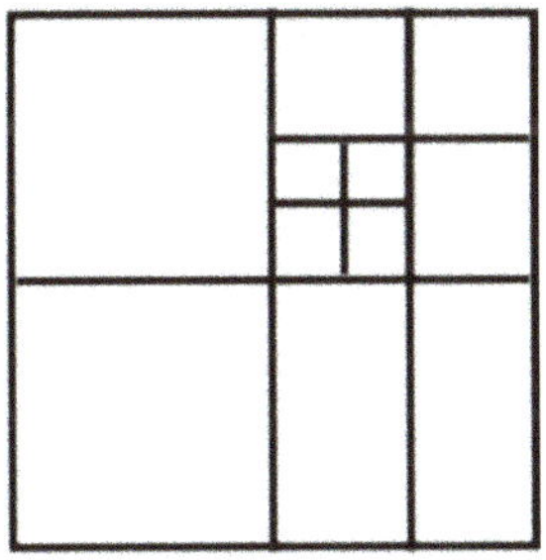

Figure 3.3: Nestedness property.

We shall need the following version of the Lebesgue differentiation theorem.

Lemma 3.9. *Assume that $x \in \mathbb{R}^n$ is a Lebesgue point of $f \in L^1_{\mathrm{loc}}(\mathbb{R}^n)$. Then*

$$f(x) = \lim_{k \to +\infty} \frac{1}{m(Q_k)} \int_{Q_k} f(y)\, dy,$$

whenever $Q_1, Q_2, Q_3, \ldots$ is any sequence of cubes containing x such that

$$\lim_{k \to +\infty} m(Q_k) = 0.$$

The Lebesgue differentiation theorem not only holds for balls but also for cubes and dyadic cubes.

Proof. Letting $Q_k = Q(x_k, l_k)$, where $x_k \in \mathbb{R}^n$ is the center and $l_k = l(Q_k)$ is the side length of the cube Q_k for every $k = 1, 2, \ldots$, we observe that $Q(x_k, l_k) \subseteq B(x, \sqrt{n} l_k)$ (see Figure 3.4) for $k = 1, 2, \ldots$. This implies

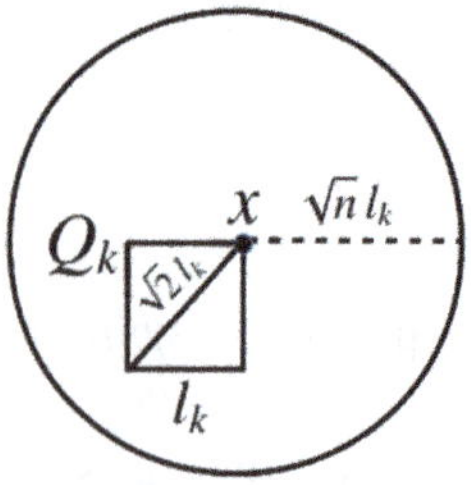

Figure 3.4: Illustration of $Q(x_k, l_k) \subseteq B(x, \sqrt{n}l_k)$.

$$\begin{aligned}\left|\fint_{Q(x_k,l_k)} f(y)\,dy - f(x)\right| &\le \fint_{Q(x_k,l_k)} |f(y) - f(x)|\,dy \\ &\le \frac{m(B(x,\sqrt{n}l_k))}{m(Q(x_k,l_k))} \fint_{B(x,\sqrt{n}l_k)} |f(y) - f(x)|\,dy \\ &= m(B(0,1))n^{n/2} \fint_{B(x,\sqrt{n}l_k)} |f(y) - f(x)|\,dy \to 0\end{aligned}$$

when $k \to +\infty$, since $l_k \to 0$ as $k \to +\infty$. □

Theorem 3.17 (Calderón–Zygmund decomposition of a cube). *Assume that $f \in L^1_{\mathrm{loc}}(\mathbb{R}^n)$ and let Q be a cube in $\mathbb{R}^n$. Then for every $t \ge \fint_Q |f(y)|\,dy$, there are countably or finitely many dyadic subcubes Q_k, $k = 1, 2, \ldots$ of Q such that:*

(1) *the interiors of Q_k, $k = 1, 2, \ldots$ are pairwise disjoint,*

(2) *$t < \fint_{Q_k} |f(y)|\,dy \le 2^n t$ for every $k = 1, 2, \ldots$,*

(3) *$|f(x)| \le t$ for almost every $x \in Q \setminus \bigcup_{k=1}^{+\infty} Q_k$.*

The collection of cubes Q_k, $k = 1, 2, \ldots$ is called the Calderón–Zygmund cubes in Q at level t.

Proof. The strategy of the proof is the following stopping time argument. For every $x \in Q$ such that $|f(x)| > t$, we choose the large dyadic cube $Q' \in \mathcal{D}$ containing x such that $\fint_{Q'} |f(y)|\,dy > t$.

Then we use the fact that for any collection of dyadic subcubes of Q, there is a subcollection of dyadic cubes with disjoint interiors and with the same union as the original cubes. These are the desired Calderón–Zygmund cubes.

Now we give a rigorous argument by considering (possibly empty) collection $\mathcal{D}'$ of dyadic subcubes $Q' \in \mathcal{D}$ of Q that satisfy

$$\fint_{Q'} |f(y)|\,dy > t. \tag{3.13}$$

The cubes in $\mathcal{D}'$ are not necessarily pairwise disjoint. Nevertheless, we can construct a new collection of cubes such that for every $Q' \in \mathcal{D}'$ we consider all cubes $Q' \in \mathcal{D}'$

such that $Q' \subseteq Q''$. Since $\fint_Q |f(y)\,dy \leq t$, for every cube $Q' \in \mathcal{Q}'$ there exists a maximal cube $Q'' \in \mathcal{D}'$. Let $\mathcal{D} = \{Q_k\}_k$ be the collection of these maximal cubes with respect to inclusion.

We show that this collection has the desired properties. Indeed,

- Property (1) follows immediately from the maximality of the cubes in $\mathcal{D}$ and the nestedness property of the dyadic subcubes.
  Indeed, if the interiors of two different cubes in $\mathcal{D}$ intersect then one is contained in the other and hence one of them cannot be maximal, see Figure 3.5.

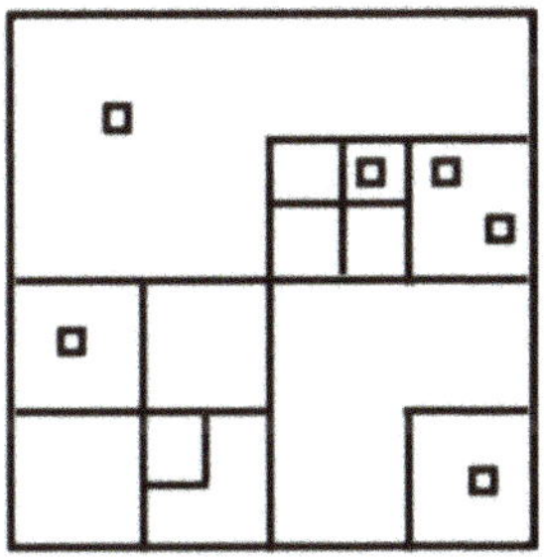

Figure 3.5: Collection of maximal subcubes.

- By (3.12), we note that $Q \in \mathcal{D}$. If $Q_k \in \mathcal{D} \cap \mathcal{D}_k$ for some k, then by properties 4 and 5 of the dyadic subcubes, we conclude that Q_k is contained in some cube $Q' \in \mathcal{D}_{k-1}$ with $m(Q') = 2^n\, m(Q_k)$; see Figure 3.6.

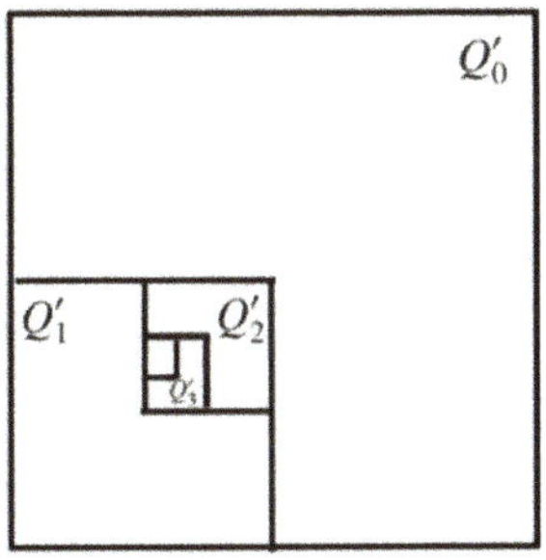

Figure 3.6: Illustration of $Q'_k \in \mathcal{D}$ such that $x \in Q'_k$ for every $k = 1, 2, \ldots$.

Since Q_k maximal, cube Q' does not satisfy (3.13). Thus

$$t < \frac{1}{m(Q_k)} \int_{Q_k} |f(y)|\,dy \leq \frac{m(Q')}{m(Q_k)} \frac{1}{m(Q')} \int_{Q_k} |f(y)|\,dy \leq 2^n\, t.$$

- Assume that $x \in Q \setminus \bigcup_{k=1}^{+\infty} Q_k$. As in the beginning of the proof, $t \geq \fint_{Q'} |f(y)|\,dy$ for every dyadic subcube $Q' \in \mathcal{D}$ containing point x. Thus there exists $Q'_k \in \mathcal{D}_k$ such that $x \in Q'_k$ for every $k = 1, 2, \ldots$.
  Note that $Q'_1 \supset Q'_2 \supset \cdots$ and $\bigcap_{k=1}^{+\infty} Q'_k = \{x\}$, see Figure 3.6. If x is a Lebesgue point of f, Lemma 3.9 implies

$$|f(x)| = \lim_{k\to+\infty} \frac{1}{m(Q'_k)} \int_{Q'_k} |f(y)|\,dy \leq t. \qquad \square$$

4 Muckenhoupt weights and the A_p condition

Weighted function spaces play a crucial role in modern analysis, particularly in the study of singular integrals, maximal functions, and PDEs. Among the most important classes of weights are the *Muckenhoupt A_p weights*, which characterize the boundedness of the Hardy–Littlewood maximal operator and other fundamental operators on weighted Lebesgue spaces. These weights arise naturally in questions of harmonic analysis, interpolation theory, and applications to partial differential equations, making them indispensable in both theoretical and applied contexts.

Most of the results of previous chapters hold if we change the domain of integration, namely $(0, +\infty)$, to $\mathbb{R}$ or even to $\mathbb{R}^n$. So, in this section, we will work on $\mathbb{R}^n$, and, as an exercise, the reader is invited to provide the corresponding details.

The question we raise is whether there is a characterization off all weights $\omega(x)$ such that the strong type (p, p) inequality

$$\int_{\mathbb{R}^n} (M(f)(x))^p \omega(x)\, dx \le C_p \int_{\mathbb{R}^n} |f(x)|^p\, \omega(x)\, dx \tag{4.1}$$

is valid for all $f \in L_p(\omega)$.

Suppose that (4.1) is valid for some weight ω and all $f \in L_p(\omega)$ for some $1 < p < +\infty$. Apply (4.1) to the function $f\chi_B$ supported in a ball B and use the fact that

$$Mf\, \mathcal{X}_B(x) \ge \frac{1}{m(B)} \int_B |f(y)|\, dy,$$

for all $x \in B$, to obtain

$$\omega(B)\left[\frac{1}{m(B)} \int_B |f(y)|\, dy\right]^p \le \int_B [Mf\, \mathcal{X}_B(x)]^p\, \omega(x)\, dx \le C_p \int_B |f(y)|^p\, \omega(x)\, dx,$$

from which it follows that

$$\left[\frac{1}{m(B)} \int_B |f(y)|\, dy\right]^p \le \frac{C_p}{\omega(B)} \int_B |f(y)|^p\, \omega(x)\, dx \tag{4.2}$$

for all balls B and all functions f. At this point, it is tempting to choose a function such that the two integrands are equal.

We do so by setting $f = \omega^{-q/p}$ where $\frac{1}{p} + \frac{1}{q} = 1$, which gives

$$f^p \omega = (\omega^{-q/p})^p \omega = \omega^{1-q} = \omega^{-q/p}.$$

Under the assumption that $\inf_B \omega > 0$ for all balls B, it would follow from (4.2) that

https://doi.org/10.1515/9783112223246-004

$$\left[\frac{1}{m(B)}\int_B (\omega(x))^{-q/p}\,dx\right]^p \le \frac{C_p^p}{\omega(B)}\int_B (\omega(x))^{-q/p}\,dx$$

and thus

$$\omega(B)\left[\frac{1}{m(B)}\int_B (\omega(x))^{-q/p}\,dx\right]^{p-1} \le C_p^p.$$

Therefore,

$$\sup_B\left[\frac{1}{m(B)}\int_B \omega(x)\,dx\right]\left[\frac{1}{m(B)}\int_B (\omega(x))^{-\frac{1}{p-1}}\,dx\right]^{p-1} \le C_p^p. \tag{4.3}$$

If $\inf_B \omega = 0$ for some ball B, we take $f = (\omega+\epsilon)^{-\frac{q}{p}}$ to obtain

$$\left[\frac{1}{m(B)}\int_B (\omega(x)\,dx\right]\left[\frac{1}{m(B)}\int_B (\omega(x)+\epsilon)^{-\frac{1}{p-1}}\,dx\right]^{p-1} \le C_p^p, \tag{4.4}$$

from which we deduce (4.3) via the Lebesgue monotone convergence theorem by letting $\epsilon \to 0$.

We have now obtained that every weight ω satisfying (4.1) must also satisfy the rather strange-looking condition (4.3), which we refer to in the sequel as the A_p condition.

Observe that if (4.2) is true, then (4.2) is nothing more then Hölder's (or Jensen's) inequality with Lebesgue measure on the right-hand side and the weighted measure on the left-hand side. In particular, this implies that every function in $L_p(\omega)$ is locally integrable.

Definition 4.1. Let $1 < p < +\infty$. A weight ω is said to be of class A_p if

$$\sup_{B \text{ ball in } \mathbb{R}^n}\left[\frac{1}{m(B)}\int_B (\omega(x)\,dx\right]\left[\frac{1}{m(B)}\int_B (\omega(x))^{-\frac{1}{p-1}}\,dx\right]^{p-1} < +\infty. \tag{4.5}$$

The expression in (4.5) is called the A_p Muckenhoupt characteristic constant of ω and will be denoted by $[\omega]_p$.

Remark 4.1. The A_p condition first appeared in a somewhat different form in a paper by M. Rosenblum (see [61]). The characterization of A_p when $n = 1$ is due to B. Muckenhoupt (see [58]).

The next result deals with the so called Vitali covering of a set (see Figure 4.1).

Theorem 4.1 (Vitali covering theorem). *Let $E \subset \mathbb{R}^n$ be a bounded set. Assume that $\mathcal{F}$ is a collection of open balls which are centered at points of E such that every point of E is the center of some ball in $\mathcal{F}$. Then there exists a sequence (possibly terminating) $B_1, B_2, \ldots$ of balls from $\mathcal{F}$ such that*

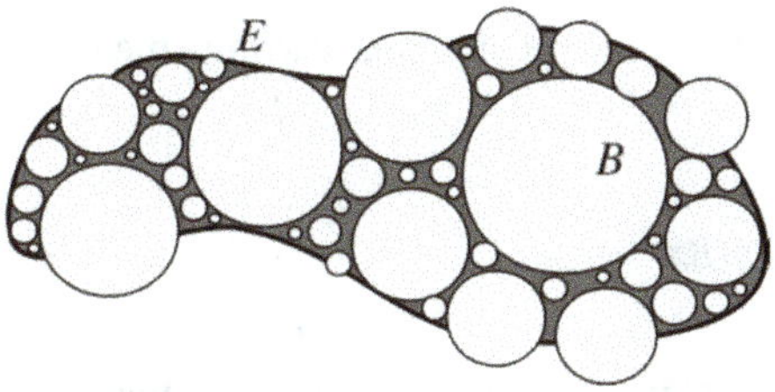

Figure 4.1: Vitali covering.

(1) *The balls $B_1, B_2, \ldots$ are disjoint,*
(2) $E \subseteq \bigcup_{\alpha \geq 1} 3B_\alpha$.

Notice that E is not covered by the disjoint balls, but is covered by the concentric balls of triple radius.

Proof. This proof is actually attributed to Stefan Banach. Its based on a very simple idea: select the balls inductively, at each stage choosing the largest one which is disjoint from the previous selections. This recipe has to be modified, however, since there may be infinitely many radii involved and there may not be a largest radius available. We shall just be sure to select a ball whose radius is almost as large as possible.

Before starting the selection, we observe that if the radii of the balls in $\mathcal{F}$ are not bounded above, then since E is bounded we can choose a single ball $B \in \mathcal{F}$ whose radius is so large (and, of course, centered at a point of E) that $E \subset B$. Therefore, we assume that the radii of the balls in $\mathcal{F}$ are bounded above. We select the balls $B_1, B_2, \ldots$ inductively in the following manner: Assume that $B_1, B_2, \ldots, B_{\alpha-1}$ have been selected, where $\alpha \geq 1$. Define

$$d_\alpha := \sup\left\{\operatorname{rad}(B) : B \in \mathcal{F} \text{ and } B \cap \bigcup_{\beta<\alpha} B_\beta = \emptyset\right\},$$

where $\operatorname{rad}(B)$ is radius of ball B.

If there is no $B \in \mathcal{F}$ satisfying $B \cap \bigcap_{\beta<\alpha} B_\beta = \emptyset$, the process terminates with $B_{\alpha-1}$; otherwise, we choose $B_\alpha \in \mathcal{F}$ such that $\frac{1}{2}d_\alpha < \operatorname{rad}(B_\alpha)$ and $B_\alpha \cap \bigcap_{\beta<\alpha} B_\beta = \emptyset$.

This scheme not only serves to select the balls inductively, but also in the case $\alpha = 1$ gives the method for selecting the first ball B_1.

The selection certainly makes sense as long as it can be performed, which is true since our assumptions guarantee that $0 < d_\alpha < +\infty$. Of the two properties the balls are required to have, (1) is clear.

In order to prove (2), let $x \in E$ be arbitrary. There exists a $B \in \mathcal{F}$ having center x. Let $\rho = \operatorname{rad}(B)$. We first notice that B must have a nonempty intersection with at least one of the selected balls $B_1, B_2, \ldots$. Otherwise, $B \cap B_\alpha = \emptyset$ for all α. This implies that the selection never terminates, and indeed that $\rho \leq d_\alpha$ for $\alpha = 1, 2, \ldots$.

This in turn implies that we have countably many balls B_α which are disjoint, whose centers lie in a bounded set E and whose radii satisfy $\operatorname{rad}(B_\alpha) > \frac{1}{2}d_\alpha \geq \frac{1}{2}\rho > 0$.

This is clearly impossible since $\bigcup_{\alpha=1}^{+\infty} B_\alpha$ is a bounded set and thus has finite measure, yet we get

$$m\left(\bigcup_{\alpha=1}^{+\infty} B_\alpha\right) = \sum_{\alpha=1}^{+\infty} m(B_\alpha) = +\infty.$$

Since B meets at least one B_α, there exist a smallest $\alpha \geq 1$ such that $B \cap B_\alpha \neq \emptyset$. Therefore, $B \cap \bigcup_{\beta=1}^{\alpha-1} B_\beta = \emptyset$ and conclude

$$\rho \leq d_\alpha < 2\,\mathrm{rad}(B_\alpha).$$

Let $y \in B \cap B_\alpha$. If z is the center of B_α, then

$$\begin{aligned}|x-z| &\leq |x-y| + |y-z| \\ &< \rho + \mathrm{rad}(B_\alpha) \\ &< 3\,\mathrm{rad}(B_\alpha).\end{aligned}$$

Therefore, $x \in 3B_\alpha$, and so $E \subset \bigcup_{\alpha\geq 1} 3B_\alpha$. □

The factor 3 which appears in conclusion (2) of the theorem is not optimal. In fact, a trivial modification of the above proof shows that this factor can be replaced by $2+\epsilon$ for any $\epsilon > 0$. All that is necessary is to choose $B_\alpha \in \mathcal{F}$ to satisfy

$$\frac{1}{1+\epsilon} d\alpha < \mathrm{rad}(B_\alpha).$$

The factor cannot be reduced further, in the sense that it cannot be guaranteed that $E \subset \bigcup_{\alpha\geq 1} 2B_\alpha$. For instance, let $E = (-1,1)$. Take $x \in E$. Let $r(x) = \frac{1+2|x|}{3}$ and $B_x = (x - r(x), x + r(x))$. Consider $\mathcal{F} = \{B_x : x \in E\}$. It is not hard to see that the conclusion of Vitali theorem does not hold if the factor 3 is replaced by 2.

Proposition 4.1. *Let $f \in L_p(\omega)$, $1 < p < +\infty$. Then $Mf \in L_p(\omega)$. Moreover,*

$$\|Mf\|_{L_p(\omega)} \leq [\omega]_p^{\frac{1}{p}} \|f\|_{L_p(\omega)}.$$

Proof. Let $f \in L_p(\omega)$ with $1 < p < +\infty$. Then by Hölder's inequality, we have

$$\begin{aligned}|Mf(x)|^p &= \left(\frac{1}{m(B)} \int_B |f(x)|\,dx\right)^p \\ &= \left(\frac{1}{m(B)} \int_B |f(x)|\,(\omega(x))^{\frac{1}{p}}\,(\omega(x))^{-\frac{1}{p}}\,dx\right)^p \\ &\leq \left(\frac{1}{m(B)}\right)^p \left(\int_B |f(x)|^p\,\omega(x)\,dx\right)\left(\int_B (\omega(x))^{-\frac{q}{p}}\,dx\right)^{\frac{p}{q}}.\end{aligned} \tag{4.6}$$

Next, integrating both sides of (4.6) over B, we have

$$
\begin{aligned}
&\int_B |Mf(x)|^p \omega(x)\,dx \\
&\quad \le \left(\int_B |f(x)|^p \omega(x)\,dx\right)\left(\frac{1}{m(B)}\int_B (\omega(x))\,dx\right)\left(\frac{1}{m(B)}\int_B (\omega(x))^{-\frac{1}{p-1}}\,dx\right)^{p-1} \\
&\quad \le [\omega]_p \int_B |f(x)|^p \omega(x)\,dx,
\end{aligned}
$$

and thus

$$
\|Mf\|_{L_p(\omega)} \le [\omega]_p^{\frac{1}{p}} \|f\|_{L_p(\omega)}. \qquad \square
$$

Now, let us recall the definition of weighted Hardy–Littlewood maximal function on $\mathbb{R}^n$ over balls

$$
M_\omega(f)(x) = \sup_{x\in B} \frac{1}{m(B)} \int_B |f(y)|\,\omega(y)\,dy,
$$

where ω is any weight.

In the following theorem, our proof avoids the Calderón–Zygmund decomposition (see [4]). Instead, we use Theorem 4.1 (Vitali covering theorem) and the fact that ω as a measure satisfies the doubling condition, i. e.,

$$
\omega(\lambda B) \le \lambda^{np}\,[\omega]_p\,\omega(B);
$$

see also [21].

Theorem 4.2. *For $1 \le p < +\infty$, the weak (p,p) inequality*

$$
\omega(\{x \in \mathbb{R}^n : Mf(x) > \lambda\}) \le \frac{C}{\lambda^p} \int_{\mathbb{R}^n} |f(x)|^p\,\omega(x)\,dx \quad \textit{holds if } \omega \in A_p.
$$

Proof. In the proof of Proposition 4.1, we got that

$$
\left(\frac{1}{m(B)}\int_B |f(x)|\,dx\right)^p \le \left(\frac{1}{m(B)}\right)^p \left(\int_B |f(x)|^p \omega(x)\,dx\right)\left(\int_B (\omega(x))^{-\frac{q}{p}}\,dx\right)^{\frac{p}{q}}.
$$

The right-hand side term in the above inequality is

$$
\begin{aligned}
&\left(\frac{1}{m(B)}\right)^p \left(\int_B |f(x)|^p \omega(x)\,dx\right)\left(\int_B (\omega(x))^{-\frac{q}{p}}\,dx\right)^{\frac{p}{q}} \\
&\quad \le \left(\frac{1}{m(B)}\int_B |f(x)|^p \omega(x)\,dx\right)\left(\frac{\omega(B)}{m(B)}\right)\left(\frac{1}{m(B)}\int_B (\omega(x))^{-\frac{1}{p-1}}\,dx\right)^{p-1} \\
&\quad \le \left(\frac{1}{m(B)}\int_B |f(x)|^p \omega(x)\,dx\right)[\omega]_p,
\end{aligned}
$$

and thus

$$\left(\frac{1}{m(B)}\int_B |f(x)|\,dx\right)^p \leq [\omega]_p\left(\frac{1}{m(B)}\int_B |f(x)|^p\,\omega(x)\,dx\right). \tag{4.7}$$

Since

$$M_\omega(f)(x) = \sup_{x\in B}\frac{1}{m(B)}\int_B |f(y)|\,\omega(y)\,dy,$$

fixing $\lambda > 0$, from (4.7) we get

$$\{x \in \mathbb{R}^n : M(f)(x) > \lambda\} \subset \left\{x \in \mathbb{R}^n : M_\omega(f)(x) > \frac{\lambda^p}{[\omega]_p}\right\}.$$

Thus

$$\omega(\{x \in \mathbb{R}^n : M(f)(x) > \lambda\}) \leq \omega\left(\left\{x \in \mathbb{R}^n : M_\omega(f)(x) > \frac{\lambda^p}{[\omega]_p}\right\}\right).$$

Consider $A_\lambda \cap B(0,k)$ (with k fixed), where

$$A_\lambda = \left\{x \in \mathbb{R}^n : M_\omega(f)(x) > \frac{\lambda^p}{[\omega]_p}\right\}.$$

We assume $A_\lambda \neq \emptyset$, of course, since result is trivial otherwise. For each $x \in A_\lambda$, there exits an $r > 0$ (depending on x) such that

$$\frac{\lambda^p}{[\omega]_p} < \frac{1}{\omega(B_r)}\int_{B_r} |f(x)|^p\,\omega(x)\,dx. \tag{4.8}$$

After obtaining an estimate for the measure of $A_\lambda \cap B(0,k)$, we can let $k \to +\infty$.

Now, let **F** be the collection of open balls B with centers in $A_\lambda \cap B(0,k)$ and satisfying (4.8). Then the hypotheses of Theorem 4.1 (Vitali covering theorem) are satisfied. However, if $A_\lambda \cap B(0,k) \neq \emptyset$, then there exist balls $B_1, B_2, \ldots \in$ **F** such that

i) $B_1, B_2, \ldots$ are disjoint,
ii) $A_\lambda \cap B(0,k) \subset \bigcup_{r\geq 1} 3B_r$.

All we have to do is assemble this information. Here is the method: first we use the inclusion (4.2), then the fact that

$$\omega(3B_r) \leq 3^{np}\,[\omega]_p\,\omega(B_r),$$

due to the inequality (4.8), (4.1), and the disjointedness of the selected balls. We obtain

$$
\begin{aligned}
\omega(A_\lambda \cap B(0,k)) &\le \sum_{r\ge 1} \omega(3B_r) \le 3^{np}\,[\omega]_p \sum_{r\ge 1} \omega(B_r)\\
&\le 3^{np}\,\frac{[\omega]_p^2}{\lambda^p} \sum_{r\ge 1} \int_{B_r} |f(x)|^p \omega(x)\,dx\\
&= 3^{np}\,\frac{[\omega]_p^2}{\lambda^p} \int_{\bigcup_{r\le 1} B_r} |f(x)|^p \omega(x)\,dx\\
&= 3^{np}\,\frac{[\omega]_p^2}{\lambda^p} \int_{\mathbb{R}^n} |f(x)|^p \omega(x)\,dx.
\end{aligned}
$$

Finally, we let $k \to +\infty$ to obtain

$$
\omega(A_\lambda) \le 3^{np}\,\frac{[\omega]_p^2}{\lambda^p} \int_{\mathbb{R}^n} |f(x)|^p \omega(x)\,dx,
$$

that is,

$$
\omega(\{x \in \mathbb{R}^n : M(f)(x) > \lambda\}) \le 3^{np}\,\frac{[\omega]_p^2}{\lambda^p} \int_{\mathbb{R}^n} |f(x)|^p \omega(x)\,dx. \qquad \square
$$

We can conclude certain elementary properties of weights directly from (4.8).

Lemma 4.1. *The following properties hold for a nonnegative Lebesgue measurable function ω that satisfies (4.8):*

1. *Either $\omega = 0$ or $\omega > 0$ almost everywhere in $\mathbb{R}^n$.*
2. *There is a constant c such that $\omega(2B) \le c\,\omega(B)$ for every ball $B \subset \mathbb{R}^n$.*
   In this case, we say that ω is a doubling measure.
3. *Either $\omega \in L^1_{\mathrm{loc}}(\mathbb{R}^n)$ or $\omega = +\infty$ almost everywhere in $\mathbb{R}^n$.*

In order to avoid trivialities, we assume that $\omega > 0$ almost everywhere in $\mathbb{R}^n$ and that $\omega \in L^1_{\mathrm{loc}}(\mathbb{R}^n)$. In this case, the Lebesgue measure and the weighted measure have the same classes of measurable sets and measurable functions.

Proof. 1. Assume that there is a set of positive measure on which the weight is zero. In other words, let $A = \{x \in \mathbb{R}^n : \omega(x) = 0\}$ and assume that $m(A) > 0$. We begin by showing that A can be assumed bounded. Since

$$
0 < m(A) = m\Big(\bigcup_{k=1}^{+\infty} (B(0;k) \cap A)\Big) = \lim_{k\to+\infty} m(B(0;k) \cap A),
$$

by choosing k_0 large enough, we have $m(Q(0;k_0) \cap A) > 0$. For $k \ge k_0$, the estimate (4.8) gives

$$\begin{aligned}\int_{B(0;k)} \omega(x)\,dx = \omega(B(0;k)) &\le C\left(\frac{m(B(0;k))}{m(B(0;k_0)\cap A)}\right)^p \omega(B(0;k_0)\cap A)\\ &\le C\left(\frac{m(B(0;k))}{m(B(0;k_0)\cap A)}\right)^p \int_A \omega(x)\,dx\\ &= 0.\end{aligned}$$

It follows that

$$\int_{\mathbb{R}^n} \omega(x)\,dx = \int_{\bigcup_{k=k_0}^{+\infty} B(0;k)} \omega(x)\,dx = \lim_{k\to+\infty} \int_{B(0;k)} \omega(x)\,dx = 0.$$

Since $\omega \ge 0$, we conclude that $\omega = 0$ almost everywhere in $\mathbb{R}^n$.

2. By replacing B with $2B$ and A with B in (4.8), we have

$$\omega(2B)\left(\frac{m(B)}{m(2B)}\right)^p \le C\omega(B),$$

which implies $\omega(2B) \le C2^{np}\omega(B)$ for every ball $B \subseteq \mathbb{R}^n$.

3. Assume that $m(\{x \in \mathbb{R}^n : \omega(x) < +\infty\}) > 0$. Then

$$\begin{aligned}0 < m(\{x \in \mathbb{R}^n : \omega(x) < +\infty\}) &= m\left(\bigcup_{k=1}^{+\infty}\{x \in B(0;k) : \omega(x) < +\infty\}\right)\\ &= \lim_{k\to+\infty} m(\{x \in B(0;k) : \omega(x) < +\infty\}).\end{aligned}$$

Thus there exists a k_0 such that $m(\{x \in B(0;k_0) : \omega(x) < +\infty\}) > 0$ and then

$$\begin{aligned}0 < m(\{x \in B(0;k_0) : \omega(x) < +\infty\}) &= m\left(\bigcup_{j=1}^{+\infty}\{x \in B(0;k_0) : \omega(x) < j\}\right)\\ &= \lim_{j\to+\infty} m(\{x \in B(0;k_0) : \omega(x) < j\}).\end{aligned}$$

Thus there exists a j_0 such that $m(\{x \in B(0;k_0) : \omega(x) < j_0\}) > 0$.
Denote $A = m(\{x \in B(0;k_0) : \omega(x) < j_0\})$ and recall $m(A) > 0$. For $k \ge k_0$, the estimate (4.8) gives

$$\begin{aligned}\int_{B(0;k)} \omega(x)\,dx = \omega(B(0;k)) &\le C\left(\frac{m(B(0:k))}{m(A)}\right)^p \omega(A)\\ &= C\left(\frac{m(B(0:k))}{m(A)}\right)^p \omega(\{x \in B(0;k_0) : \omega(x) < j_0\})\\ &= C\left(\frac{m(B(0:k))}{m(A)}\right)^p \int_{\{x\in B(0;k_0):\omega(x)<j_0\}} \omega(x)\,dx\\ &\le C\left(\frac{m(B(0:k))}{m(A)}\right)^p j_0\omega(\{x \in B(0;k_0) : \omega(x) < j_0\})\end{aligned}$$

$$= C\Big(\frac{m(B(0:k))}{m(A)}\Big)^p j_0 \omega(B(0;k_0))$$

$$< +\infty.$$

Thus $\omega \in L^1_{\mathrm{loc}}(B(0;k))$ for every $k \le k_0$ and, consequently, $\omega \in L^1_{\mathrm{loc}}(\mathbb{R}^n)$.

Now, let us consider the case $p = 1$. If $m(A) > 0$, then (4.8) gives

$$\frac{1}{m(B)}\int_B \omega(x)\,dx \le C\,\frac{\omega(A)}{m(A)}.$$

Denote $a = \operatorname{ess\,inf}_{x\in B}\omega(x)$ and let $\epsilon > 0$. Recall that

$$\operatorname*{ess\,inf}_{x\in B}\omega(x) = \sup\{M \in \mathbb{R} : \omega(x) > M \text{ for almost every } x \in B\}.$$

Then there exist $A_\epsilon \subset B$ such that $m(A_\epsilon) > 0$ and $\omega(x) < a + \epsilon$ for every $x \in A_\epsilon$. Thus

$$\frac{1}{m(B)}\int_B \omega(x)\,dx \le C\,\frac{1}{m(A_\epsilon)}\int_{A_\epsilon}(a+\epsilon)\,dx = C(a+\epsilon) = C\,(\operatorname{ess\,inf}\omega(x) + \epsilon),$$

from which it follows that

$$\frac{1}{m(B)}\int_B \omega(x)\,dx \le C\,\operatorname*{ess\,inf}_{x\in B}\omega(x).$$

This leads to the definition of the Muckenhoupt class A_1. □

Definition 4.2. A weight ω for which there exists a constant C, independent of the ball B, such that

$$\frac{1}{m(B)}\int_B \omega(x)\,dx \le C\operatorname*{ess\,inf}_{x\in B}\omega(x), \tag{4.9}$$

for every ball $B \subseteq \mathbb{R}^n$, is called an A_1 weight. The smallest constant C for which (4.9) holds is called the A_1 constant of ω.

Equivalently, (4.9) can be written in the form

$$\frac{1}{m(B)}\int_B \omega(x)\,dx \le C\,\omega(x) \quad \text{for almost every } x \in B. \tag{4.10}$$

By the Lebesgue differentiation theorem, $\omega(x) \le M\,\omega(x) \le C\,\omega(x)$ for almost every $x \in \mathbb{R}^n$. Thus, an A_1 weight is pointwise comparable to its maximal function. This is a maximal function characterization of A_1.

Theorem 4.3. *A weight $\omega \in A_1$ if and only if*

$$M\,\omega(x) \le C\,\omega(x) \quad \textit{for almost every } x \in \mathbb{R}^n. \tag{4.11}$$

Proof. ($\Longleftarrow$) It is clear that (4.11) implies (4.10), since

$$\frac{1}{m(B)}\int_B \omega(y)\,dy \le M\,\omega(x) \le C\,\omega(x) \quad \text{for almost every } x \in \mathbb{R}^n.$$

($\Longrightarrow$) Assume that $\omega \in A_1$. We claim that $m(\{x \in \mathbb{R}^n : M\,\omega(x) > C\,\omega(x)\}) = 0$.
If $x \in \mathbb{R}^n$ with $M\,\omega(x) > C\,\omega(x)$, then there exists a ball B containing x such that

$$\frac{1}{m(B)}\int_B \omega(y)\,dy \ge C\,\omega(x). \tag{4.12}$$

For every $\epsilon > 0$, there is a ball $\widetilde{B}$ whose boundary has rational coordinates with $B \subseteq \widetilde{B}$ and $m(\widetilde{B} \setminus B) < \epsilon$. Note that $m(\widetilde{B}) = m(B) + m(\widetilde{B} \setminus B) < m(B) + \epsilon$. By choosing $\epsilon > 0$ small enough, we have

$$\frac{1}{m(\widetilde{B})}\int_{\widetilde{B}} \omega(y)\,dy \ge \frac{1}{m(B)+\epsilon}\int_B \omega(y)\,dy > C\,\omega(x).$$

Thus we may assume that the boundary points of the ball B satisfying (4.12) are rational. The A_1 condition and (4.12) imply

$$C\,\omega(x) < \frac{1}{m(B)}\int_B \omega(y)\,dy \le C \operatorname*{ess\,inf}_{y\in B} \omega(y),$$

and thus

$$\omega(x) < \operatorname*{ess\,inf}_{y\in B} \omega(y).$$

This implies that $x \in A$, $A \subseteq B$, with $m(A) = 0$. Since there are at most countably many balls B_i, $i = 1, 2, \ldots$ with rational boundaries satisfying (4.12), we have

$$\{x \in \mathbb{R}^n : M\,\omega(x) > C\,\omega(x)\} \subset \bigcup_{i=1}^{+\infty} (B_i \cap A_i),$$

where $A_i \subseteq B_i$ with $m(A_i) = 0$, $i = 1, 2, \ldots$. This implies

$$m(\{x \in \mathbb{R}^n : M\,\omega(x) > C\,\omega(x)\}) \le \sum_{i=1}^{+\infty} m(B_i \cap A_i) \le \sum_{i=1}^{+\infty} m(A_i) = 0. \qquad \square$$

4.1 Properties of A_p weights

It is straightforward that translations, isotropic dilations, and scalar multiples of A_p weights are also A_p weights with the same A_p characteristic constant.

We summarize some basic properties of A_p weights in the following proposition.

Proposition 4.2. *Let $\omega \in A_p$ for some $1 \le p < +\infty$. Then*
1. $[\delta^\lambda(\omega)]_p = [\omega]_p$ *for all* $\lambda > 0$.
2. $[\tau^z(\omega)]_p = [\omega]_p$, *where* $\tau^z(\omega)(z) = \omega(x-z)$, $z \in \mathbb{R}^n$.
3. $[\lambda\omega]_p = [\omega]_p$ *for all* $\lambda > 0$.
4. $[\omega]_p \ge 1$ *for all* $\omega \in A_p$. *The equality holds if and if* ω *is a constant.*
5. *For* $1 \le p < q < +\infty$, *we have* $[\omega]_p \le [\omega]_q$.
6. $\lim_{p\to 1^+}[\omega]_p = [\omega]_1$ *if* $\omega \in A_1$.
7. *The following is an equivalent characterization of the* A_p *characteristic constant of* ω:

$$[\omega]_p = \sup_B \sup_{\substack{f\in L_p(\omega),\\ m(B\cap\{|f|=0\})}} \left\{ \frac{(\frac{1}{m(B)}\int_B |f(x)|^p\,dx)^p}{\frac{1}{m(B)}\int_B |f(x)|^p\omega(x)\,dx} \right\}.$$

8. *The measure* $\omega(x)dx$ *is doubling precisely: indeed, for all* $\lambda > 1$ *and balls* B, *we have*

$$\omega(\lambda B) = \lambda^{np}[\omega]_p\omega(B).$$

9. *Given* $\omega_1, \omega_2 \in A_1$, *for* $1 < p < \infty$, *one has* $\omega_1\omega_2^{1-p} \in A_p$.

Proof. 1.–3. The simple proofs of these properties are left as an exercise.

4. To prove this property, we use Hölder's inequality with exponents p and q to obtain

$$\begin{aligned}1 = \frac{1}{m(B)}\int_B dx &= \frac{1}{m(B)}\int_B \omega^{1/p}(x)\omega^{-1/p}(x)\,dx\\ &\le \frac{1}{m(B)}\left(\int_B \omega(x)\,dx\right)^{1/p}\left(\int_B \omega^{-q/p}(x)\,dx\right)^{1/q}.\end{aligned}$$

Thus $1 \le \frac{1}{m(B)}(\int_B \omega(x)\,dx)^{1/p}(\int_B \omega^{-q/p}(x)\,dx)^{1/q}$.
Then

$$\begin{aligned}1 &\le \left(\frac{1}{(m(B))^p}\int_B \omega(x)\,dx\right)\left(\int_B \omega^{-q/p}(x)\,dx\right)^{p/q}\\ &= \left(\frac{1}{(m(B))^p}\int_B \omega(x)\,dx\right)\left(\int_B \omega^{-\frac{1}{p-1}}(x)\,dx\right)^{p-1}\\ &= \left(\frac{1}{m(B)}\int_B \omega(x)\,dx\right)\left(\frac{1}{m(B)}\int_B \omega^{-\frac{1}{p-1}}(x)\,dx\right)^{p-1}\\ &\le [\omega]_p^{1/p},\end{aligned}$$

with the equality holding only when $(\omega(x))^{1/p} = C(\omega(x))^{-1/q}$ for some constant $C > 0$ (i. e., when ω is a constant).

5. If $\lambda = \frac{1-p'}{1-q'}$, where $\frac{1}{p} + \frac{1}{p'} = 1$ and $\frac{1}{q} + \frac{1}{q'} = 1$, as well as $s = \frac{1-p'}{q'-p'}$, then

$$\frac{1}{\lambda} + \frac{1}{s} = \frac{1-q'}{1-p'} + \frac{q'-p'}{1-p'} = \frac{1-p'}{1-p'} = 1.$$

Next, by Hölder's inequality,

$$\begin{aligned}\left(\int_B \omega^{1-q'}(x)\,dx\right)^{q-1} &\le ((m(B))^{1/s})^{q-1}\left(\int_B \omega^{\lambda(1-q')}(x)\,dx\right)^{\frac{q-1}{\lambda}}\\ &\le (m(B))^{\frac{q-1}{s}}\left(\int_B \omega^{1-p'}(x)\,dx\right)^{\frac{(q-1)(1-q')}{1-p'}}\\ &= (m(B))^{\frac{q-1}{s}}\left(\int_B \omega^{1-p'}(x)\,dx\right)^{-\frac{(q-1)(q'-1)}{1-q'}}.\end{aligned}$$

Since $(q-1)(q'-1) = 1$, one has

$$\left(\int_B \omega^{1-q'}(x)\,dx\right)^{q-1} \le (m(B))^{\frac{q-1}{s}}\left(\int_B \omega^{1-p'}(x)\,dx\right)^{-\frac{1}{1-p'}}. \tag{4.13}$$

Next, $\frac{1}{p}+\frac{1}{p'} = 1$ implies $\frac{1}{p} = 1-\frac{1}{p'}$, hence $\frac{1}{p'-1} = \frac{p}{p'}$. Since $\frac{p-1}{p} = \frac{1}{p'}$, we get $\frac{1}{p'-1} = p-1$. Going back to (4.13), we obtain

$$\begin{aligned}\left(\int_B \omega^{1-q'}(x)\,dx\right)^{q-1} &\le (m(B))^{\frac{q-1}{s}}\left(\int_B \omega^{1-p'}(x)\,dx\right)^{\frac{1}{p'-1}}\\ &= (m(B))^{\frac{q-1}{s}}\left(\int_B \omega^{1-p'}(x)\,dx\right)^{p-1},\end{aligned}$$

which yields

$$\begin{aligned}\frac{1}{(m(B))^q}\left(\int_B \omega^{1-q'}(x)\,dx\right)^{q-1} &\le \frac{(m(B))^{\frac{q-1}{s}}}{(m(B))^q}\left(\int_B \omega^{1-p'}(x)\,dx\right)^{p-1},\\ \frac{1}{m(B)}\left(\frac{1}{(m(B))}\int_B \omega^{1-q'}(x)\,dx\right)^{q-1} &\le (m(B))^{\frac{q-1}{s}-q}\left(\int_B \omega^{1-p'}(x)\,dx\right)^{p-1}.\end{aligned} \tag{4.14}$$

Furthermore,

$$\begin{aligned}\frac{q-1}{s} - q = \frac{q-1-sq}{s} = q\left(\frac{1-s}{s}\right) - \frac{1}{s} &= q\left(\frac{1}{s}-1\right) - \frac{1}{s}\\ &= q\left(\frac{q'-p'}{1-p'} - 1\right) - \frac{q'-p'}{1-p'}\\ &= q\left(\frac{q'-p'-1+p'}{1-p'}\right) - \frac{q'-p'}{1-p'}\\ &= \frac{1}{1-p'}(q(q'-1) - q' + p').\end{aligned} \tag{4.15}$$

Since $\frac{1}{q} + \frac{1}{q'} = 1$ implies $\frac{q'}{q} = q' - 1$, we get

$$\frac{1}{1-p'}(q(q'-1)-q'+p') = \frac{1}{1-p'}\left(q\left(\frac{q'}{q}\right)-q'+p'\right) = \frac{p'}{1-p'}.$$

But $\frac{1}{p}+\frac{1}{p'}=1$ implies $\frac{1}{p}=\frac{p'-1}{p'}$, yielding $\frac{p'}{p'-1}=p$ and $\frac{p'}{1-p'}=-p$, thus $\frac{q-1}{s}-q=-p$. Going back to (4.14), we have

$$\begin{aligned}\frac{1}{m(B)}\left(\frac{1}{(m(B))}\int_B \omega^{1-q'}(x)\,dx\right)^{q-1} &\leq \frac{1}{(m(B))^p}\left(\int_B \omega^{1-p'}(x)\,dx\right)^{p-1}\\ &= \frac{1}{(m(B))}\left(\frac{1}{(m(B))}\int_B \omega^{1-p'}(x)\,dx\right)^{p-1}.\end{aligned}$$

Finally,

$$\begin{aligned}&\frac{1}{m(B)}\int_B \omega(x)\,dx\left(\frac{1}{m(B)}\int_B \omega^{1-q'}(x)\,dx\right)^{q-1}\\ &\quad\leq \frac{1}{m(B)}\int_B \omega(x)\,dx\left(\frac{1}{(m(B))}\int_B \omega^{1-p'}(x)\,dx\right)^{p-1}.\end{aligned}$$

Thus $[\omega]_q \leq [\omega]_p$, showing that $A_p \subset A_q$.

6. For any $\epsilon > 0$ small enough, we have

$$\begin{aligned}&\left(\frac{1}{m(B)}\int_B \omega(x)\,dx\right)\left(\frac{1}{m(B)}\int_B \omega^{-\frac{1}{p-1}}(x)\,dx\right)^{p-1}\\ &\quad\leq [\omega]_p < \left(\frac{1}{m(B)}\int_B \omega(x)\,dx\right)\left(\frac{1}{m(B)}\int_B \omega^{-\frac{1}{p-1}}(x)\,dx\right)^{p-1}+\epsilon.\end{aligned}$$

Since

$$\begin{aligned}\lim_{p\to 1^+}\left(\frac{1}{m(B)}\int_B \omega^{-\frac{1}{p-1}}(x)\,dx\right)^{p-1} &= \lim_{t\to+\infty}\left(\frac{1}{m(B)}\int_B \omega^{-t}(x)\,dx\right)^{1/t}\\ &= \|\omega^{-1}\|_{L_\infty(B)},\end{aligned}$$

we obtain

$$\begin{aligned}\left(\frac{1}{m(B)}\int_B \omega(x)\,dx\right)\|\omega^{-1}\|_{L_\infty(B)} &\leq \lim_{p\to 1^+}[\omega]_p\\ &\leq \left(\frac{1}{m(B)}\int_B \omega^{-t}(x)\,dx\right)^{1/t}\|\omega^{-1}\|_{L_\infty(B)}+\epsilon.\end{aligned}$$

Hence

$$\begin{aligned}\sup_B\left(\frac{1}{m(B)}\int_B \omega(x)\,dx\right)\|\omega^{-1}\|_{L_\infty(B)} &\leq \lim_{p\to 1^+}[\omega]_p\\ &\leq \sup_B\left(\frac{1}{m(B)}\int_B \omega(x)\,dx\right)\|\omega^{-1}\|_{L_\infty(B)}.\end{aligned}$$

Thus

$$\lim_{p\to 1^+}[\omega]_p = [\omega]_1.$$

7. By Hölder's inequality,

$$\begin{aligned}
&\left(\frac{1}{m(B)}\int_B |f(x)|\,dx\right)^p \\
&= \left(\frac{1}{m(B)}\int_B |f(x)|\omega^{1/p}(x)\omega^{-1/p}(x)\,dx\right)^p \\
&\le \frac{1}{(m(B))^p}\left(\int_B |f(x)|^p\omega(x)\,dx\right)\left(\int_B \omega^{-q/p}(x)\,dx\right)^{p/q} \\
&\le \left(\frac{1}{m(B)}\int_B |f(x)|^p\omega(x)\,dx\right)\frac{\omega(B)}{m(B)}\left(\frac{1}{m(B)}\int_B \omega^{-\frac{1}{p-1}}(x)\,dx\right)^{p-1} \\
&\le \left(\frac{1}{m(B)}\int_B |f(x)|^p\omega(x)\,dx\right)[\omega]_p.
\end{aligned} \tag{4.16}$$

Conversely, let $f = (\omega+\epsilon)^{-q/p}$, $\epsilon > 0$. Then

$$\begin{aligned}
&\sup_B \sup_{\substack{f\in L_p(\omega),\\ m(B\cap\{|f|=0\})}} \left\{\frac{(\frac{1}{m(B)}\int_B |f(x)|^p\,dx)^p}{\frac{1}{\omega(B)}\int_B |f(x)|^p\omega(x)\,dx}\right\} \\
&\ge \sup_B \left\{\frac{(\frac{1}{m(B)}\int_B(\omega(x)+\epsilon)^{-q/p}\,dx)^p}{\frac{1}{\omega(B)}\int_B(\omega(x)+\epsilon)^{-q}\omega(x)\,dx}\right\} \\
&\ge \sup_B \left\{\frac{(\frac{1}{m(B)}\int_B(\omega(x))^{-q/p}\,dx)^p}{\frac{1}{\omega(B)}\int_B(\omega(x))^{-q+1}\,dx}\right\} \quad (\text{letting } \epsilon\to 0) \\
&= \sup_B \left\{\frac{\omega(B)}{m(B)}\frac{(\frac{1}{m(B)}\int_B(\omega(x))^{-\frac{1}{p-1}}\,dx)^p}{\frac{1}{\omega(B)}\int_B(\omega(x))^{-\frac{1}{p-1}}\,dx}\right\} \\
&= \sup_B \left\{\left(\frac{1}{m(B)}\int_B \omega(x)\,dx\right)\left(\frac{1}{m(B)}\int_B(\omega(x))^{-\frac{1}{p-1}}\,dx\right)^{p-1}\right\} \\
&= [\omega]_p.
\end{aligned} \tag{4.17}$$

Then, the result follows from (4.9) and (4.10).

8. To prove this property, we apply property 7 to the function $f = \mathcal{X}_B$. Taking λB in place of B, we obtain

$$\frac{(\frac{1}{m(\lambda B)}\int_B \mathcal{X}_B(x)\,dx)^p}{\frac{1}{\omega(\lambda B)}\int_B \mathcal{X}_B(x)\omega(x)\,dx} \le [\omega]_p \quad (\text{as } B\subset\lambda B),$$

$$\frac{\omega(\lambda B)(\frac{1}{m(\lambda B)}\int_B dx)^p}{\int_B \omega(x)\,dx} \le [\omega]_p.$$

Then

$$\omega(\lambda B) \le \left(\frac{m(\lambda B)}{m(B)}\right)^p \omega(B)[\omega]_p,$$

and hence

$$\omega(\lambda B) \le \lambda^{np}[\omega]_p\omega(B).$$

□

4.2 Weighted Riesz bounded variation spaces

Let $[a,b]$ be a closed interval, and let $\{x_j\}_{j=1}^n$ satisfy

$$a = x_0 < x_1 < x_2 < \cdots < x_n = b.$$

We define the finite partition $\Pi = \{x_j\}_{j=1}^n$ to be the collection of closed intervals $\pi_j = [x_{j-1}, x_j]$. For simplicity, from now on we write $\Delta f(\pi_j) = f(x_j) - f(x_{j-1})$.

This section is based on [30], where the following definition was introduced.

Definition 4.3. Consider $1 < p < \infty$, a weight ω, and an interval $[a,b]$. A function f is said to belong to the weighted Riesz bounded variation space $\mathrm{RBV}^p_\omega([a,b])$ if

$$V^p_\omega(f,[a,b]) = \sup_\pi \sum_{\pi_j \in \Pi} \left(\frac{|\Delta f(\pi_j)|}{|\pi_j|}\right)^p \omega(\pi_j) < \infty,$$

where the supremum is taken over all finite partitions Π of $[a,b]$.

We recall now the following notation. If $0 < m(E) < \infty$, where $m(E)$ is the Lebesgue measure of E, we define the average value of ω on E by

$$⨍_E \omega(x)\,dx = \frac{1}{m(E)}\int_E \omega(x)\,dx.$$

Using this notation, Definition 4.1 of class A_p may be restated as follows.

Definition 4.4. Consider $1 < p, q < \infty$, satisfying $\frac{1}{p} + \frac{1}{q} = 1$. A nonnegative locally integrable function ω is said to belong to the Muckenhoupt class A_p if $0 < \omega(x) < \infty$ a. e. (almost everywhere) and

$$[\omega]_{A_p} = \sup_Q \left(⨍_Q \omega(x)\,dx\right)\left(⨍_Q \omega(x)^{1-q}\,dx\right)^{p-1} < \infty, \tag{4.18}$$

where the supremum is taken over all cubes $Q \subset \mathbb{R}^n$.

Proposition 4.3. *If $\omega \in A_p$. Then*

$$\sup_{\Pi} \sum_{\pi_j \in \Pi} \frac{|\Delta f(\pi_j)|^p}{(\sigma(\pi_j))^{p-1}} \leq \sup_{\Pi} \sum_{\pi_j \in \Pi} \left(\frac{|\Delta f(\pi_j)|}{|\pi_j|} \right)^p \omega(\pi_j) \leq [\omega]_{A_p} \sup_{\Pi} \sum_{\pi_j \in \Pi} \frac{|\Delta f(\pi_j)|^p}{(\sigma(\pi_j))^{p-1}}, \tag{4.19}$$

where $\sigma = \omega^{1-q}$.

Proof. We shall prove first that

$$\sup_{\Pi} \sum_{\pi_j \in \Pi} \frac{|\Delta f(\pi_j)|^p}{(\sigma(\pi_j))^{p-1}} \leq \sup_{\Pi} \sum_{\pi_j \in \Pi} \left(\frac{|\Delta f(\pi_j)|}{|\pi_j|} \right)^p \omega(\pi_j).$$

For this, we will use Proposition 4.2, claim (4). We can write

$$\begin{aligned}
\sum_{\pi_j \in \Pi} \frac{|\Delta f(\pi_j)|^p}{(\sigma(\pi_j))^p} &= \sum_{\pi_j \in \Pi} \frac{|\Delta f(\pi_j)|^p}{(\int_{\pi_j} \omega^{1-q}(t)\,dt)^{p-1}} \\
&= \sum_{\pi_j \in \Pi} \frac{|\Delta f(\pi_j)|^p}{\int_{\pi_j} \omega(t)\,dt (\int_{\pi_j} \omega^{1-q}(t)\,dt)^{p-1}} \int_{\pi_j} \omega(t)\,dt \\
&= \sum_{\pi_j \in \Pi} \frac{|\Delta f(\pi_j)|^p}{|\pi_j|^p (\frac{1}{|\pi_j|} \int_{\pi_j} \omega(t)\,dt)(\frac{1}{|\pi_j|} \int_{\pi_j} \omega^{1-q}(t)\,dt)^{p-1}} \omega(\pi_j) \\
&= \sum_{\pi_j \in \Pi} \left(\frac{|\Delta f(\pi_j)|}{|\pi_j|} \right)^p \omega(\pi_j) \frac{1}{(\frac{1}{|\pi_j|} ⨍_{\pi_j} \omega(t)\,dt)(\frac{1}{|\pi_j|} ⨍_{\pi_j} \omega^{1-q}(t)\,dt)^{p-1}} \\
&\leq \sum_{\pi_j \in \Pi} \left(\frac{|\Delta f(\pi_j)|}{|\pi_j|} \right)^p \omega(\pi_j) \\
&\leq \sup_{\Pi} \sum_{\pi_j \in \Pi} \left(\frac{|\Delta f(\pi_j)|}{|\pi_j|} \right)^p \omega(\pi_j),
\end{aligned}$$

and thus

$$\sup_{\Pi} \sum_{\pi_j \in \Pi} \frac{|\Delta f(\pi_j)|^p}{(\sigma(\pi_j))^{p-1}} \leq \sum_{\pi_j \in \Pi} \left(\frac{|\Delta f(\pi_j)|}{|\pi_j|} \right)^p \omega(\pi_j). \tag{4.20}$$

Now we shall prove that

$$\sum_{\pi_j \in \Pi} \left(\frac{|\Delta f(\pi_j)|}{|\pi_j|} \right)^p \omega(\pi_j) \leq [\omega]_{A_p} \sup_{\Pi} \sum_{\pi_j \in \Pi} \frac{|\Delta f(\pi_j)|^p}{(\sigma(\pi_j))^{p-1}}.$$

Indeed,

$$\begin{aligned}
\sum_{\pi_j\in\Pi}\left(\frac{|\Delta f(\pi_j)|}{|\pi_j|}\right)^p \omega(\pi_j) &= \sum_{\pi_j\in\Pi}\left(\frac{|\Delta f(\pi_j)|}{|\pi_j|}\right)^p \int_{\pi_j}\omega(t)\,dt \\
&= \sum_{\pi_j\in\Pi}\frac{|\Delta f(\pi_j)|^p}{|\pi_j|^{p-1}}\cdot\frac{1}{|\pi_j|}\int_{\pi_j}\omega(t)\,dt \\
&= \sum_{\pi_j\in\Pi}\frac{|\Delta f(\pi_j)|^p}{|\pi_j|^{p-1}}\fint_{\pi_j}\omega(t)\,dt \\
&\le \sum_{\pi_j\in\Pi}\frac{|\Delta f(\pi_j)|^p}{|\pi_j|^{p-1}}\frac{[\omega]_{A_p}}{(\fint_{\pi_j}\omega^{1-q}(t)\,dt)^{p-1}} \\
&= \sum_{\pi_j\in\Pi}\frac{|\Delta f(\pi_j)|^p}{|\pi_j|^{p-1}}\frac{[\omega]_{A_p}}{(\frac{1}{|\pi_j|}\int_{\pi_j}\omega^{1-q}(t)\,dt)^{p-1}} \\
&= \sum_{\pi_j\in\Pi}|\Delta f(\pi_j)|^p\frac{[\omega]_{A_p}}{(\omega(\pi_j)^{1-q})^{p-1}} \\
&= [\omega]_{A_p}\sum_{\pi_j\in\Pi}\frac{|\Delta f(\pi_j)|^p}{(\sigma(\pi_j))^{p-1}} \\
&\le [\omega]_{A_p}\sup_{\Pi}\sum_{\pi_j\in\Pi}\frac{|\Delta f(\pi_j)|^p}{(\sigma(\pi_j))^{p-1}}.
\end{aligned}$$

Hence

$$\sup_{\Pi}\sum_{\pi_j\in\Pi}\left(\frac{|\Delta f(\pi_j)|}{|\pi_j|}\right)^p \omega(\pi_j) \le [\omega]_{A_p}\sup_{\Pi}\sum_{\pi_j\in\Pi}\frac{|\Delta f(\pi_j)|^p}{(\sigma(\pi_j))^{p-1}}. \tag{4.21}$$

Finally, from (4.20) and (4.20), we have (4.19) □

Proposition 4.4. *Consider $\omega\in A_p$ and $1<p<\infty$. Then*

$$\mathrm{RBV}^p_\omega([a,b]) \subset C([a,b]),$$

where $C([a,b])$ denotes the set of all continuous functions on $[a,b]$.

Proof. Given a sequence $\{x_k\}_{k\in\mathbb{N}}$ in $[a,b]$ converging to $x\in[a,b]$, by abusing notation, let $[x,x_k]$ denote the closed interval with endpoints x and x_k and $\sigma=\omega^{1-q}$. Then

$$\begin{aligned}
|f(x)-f(x_k)|^p &= \frac{|f(x)-f(x_k)|^p}{(\sigma([x,x_k]))^{p-1}}\big(\sigma([x,x_k])\big)^{p-1} \\
&\le \sup_{\Pi}\sum_{\pi_j\in\Pi}\frac{|\Delta f(\pi_j)|}{(\sigma(\pi_j))^{p-1}}\big(\sigma([x,x_k])\big)^{p-1}
\end{aligned}$$

$$\leq \left(\sup_{\Pi} \sum_{\pi_j \in \Pi} \left(\frac{|\Delta f(\pi_j)|^p}{|\pi_j|}\right)^p \omega(\pi_j)\right)(\sigma([x, x_k]))^{p-1}$$
$$= V^p_\omega(f, [a,b])(\sigma([x, x_k]))^{p-1} \to 0,$$

as $k \to \infty$.

Since this is true for every sequence, f is a continuous function on $[a,b]$. □

Though the functional V^p_ω is not a norm, if we define

$$\|f\|_{\mathrm{RBV}^p_\omega([a,b])} = |f(a)| + (V^p_\omega(f, [a,b]))^{\frac{1}{p}}, \tag{4.22}$$

then this is a norm on $\mathrm{RBV}^p_\omega([a,b])$, as we shall show in the next proposition.

Proposition 4.5. *Equation (4.22) defines a norm in* $\mathrm{RBV}^p_\omega([a,b])$.

Proof. Suppose $f = 0$ on $[a,b]$. Then it is immediate that $\|f\|_{\mathrm{RBV}^p_\omega([a,b])} = 0$.

On the other hand, if $\|f\|_{\mathrm{RBV}^p_\omega([a,b])} = 0$, then $f(a) = 0$ and for any $x \in [a,b]$, $x > a$, one has

$$0 = V^p_\omega(f, [a,b]) \geq \left(\frac{|f(a) - f(x)|^p}{|[a,x]|}\right)\omega([x,a]).$$

Therefore $f(x) = f(a) = 0$ for all $x \in [a,b]$. Thus we have proved that

$$\|f\|_{\mathrm{RBV}^p_\omega([a,b])} = 0 \quad \text{if only if} \quad f = 0.$$

Next, fix $f, g \in \mathrm{RBV}^p_\omega$. Let Π be a finite partition of $[a,b]$ and let

$$F_j = \frac{\Delta f(\pi_j)}{|\pi_j|} \quad \text{and} \quad G_j = \frac{\Delta g(\pi_j)}{|\pi_j|}.$$

Using Minkowski inequality in $L_p(\omega)$ and in the sequence space l_p, we estimate as follows:

$$\sum_{\pi_j \in \Pi} \left(\frac{|\Delta(f+g)(\pi_j)|}{|\pi_j|}\right)^p \omega(\pi_j)$$
$$= \sum_{\pi_j \in \Pi} \left(\frac{|\Delta f(\pi_j) + \Delta g(\pi_j)|}{|\pi_j|}\right)^p \omega(\pi_j)$$
$$= \sum_{\pi_j \in \Pi} \left(\frac{|\Delta f(\pi_j)|}{|\pi_j|} + \frac{|\Delta g(\pi_j)|}{|\pi_j|}\right)^p \omega(\pi_j)$$
$$= \sum_{\pi_j \in \Pi} |F_j + G_j|^p \omega(\pi_j)$$

$$
\begin{aligned}
&= \sum_{\pi_j \in \Pi} \int_{\pi_j} |F_j + G_j|^p \omega(t)\, dt \\
&= \sum_{\pi_j \in \Pi} \|F_j + G_j\|^p_{L_p(\omega)}(\pi_j) \\
&\le \sum_{\pi_j \in \Pi} (\|F_j\|^p_{L_p(\omega)}(\pi_j) + \|G_j\|^p_{L_p(\omega)}(\pi_j))^p \\
&\le \left[\left(\sum_{\pi_j \in \Pi} \|F_j\|^p_{L_p(\omega)}(\pi_j) \right)^{\frac{1}{p}} + \left(\sum_{\pi_j \in \Pi} \|G_j\|^p_{L_p(\omega)}(\pi_j) \right)^{\frac{1}{p}} \right]^p \\
&= \left[\left(\sum_{\pi_j \in \Pi} \int_{\pi_j} |F_j|^p \omega(t)\, dt \right)^{\frac{1}{p}} + \left(\sum_{\pi_j \in \Pi} \int_{\pi_j} |G_j|^p \omega(t)\, dt \right)^{\frac{1}{p}} \right]^p \\
&= \left[\left(\sum_{\pi_j \in \Pi} \left(\frac{|\Delta f(\pi_j)|}{|\pi_j|} \right)^p \omega(\pi_j) \right)^{\frac{1}{p}} + \left(\sum_{\pi_j \in \Pi} \left(\frac{\Delta g(\pi_j)}{|\pi_j|} \right)^p \omega(\pi_j) \right)^{\frac{1}{p}} \right]^p \\
&\le [(V^p_\omega(f, [a,b]))^{\frac{1}{p}} + (V^p_\omega(g, [a,b]))^{\frac{1}{p}}]^p.
\end{aligned}
$$

Thus

$$
\left(\sum_{\pi_j \in \Pi} \left(\frac{\Delta(f+g)(\pi_j)}{|\pi_j|} \right)^p \omega(\pi_j) \right)^{\frac{1}{p}} \le (V^p_\omega(f, [a,b]))^{\frac{1}{p}} + (V^p_\omega(g, [a,b]))^{\frac{1}{p}}.
$$

Hence,

$$
\begin{aligned}
&|(f+g)(a)| + \left(\sum_{\pi_j \in \Pi} \left(\frac{\Delta(f+g)(\pi_j)}{|\pi_j|} \right)^p \omega(\pi_j) \right)^{\frac{1}{p}} \\
&\quad \le |f(a)| + (V^p_\omega(f, [a,b]))^{\frac{1}{p}} + |g(a)| + (V^p_\omega(g, [a,b]))^{\frac{1}{p}} \\
&\quad \le \|f\|_{\mathrm{RBV}^p_\omega[a,b]} + \|g\|_{\mathrm{RBV}^p_\omega[a,b]},
\end{aligned}
$$

from which it follows that

$$
\|f+g\|_{\mathrm{RBV}^p_\omega([a,b])} \le \|f\|_{\mathrm{RBV}^p_\omega[a,b]} + \|g\|_{\mathrm{RBV}^p_\omega[a,b]}.
$$

Similarly, if we fix $f \in \mathrm{RBV}^p_\omega([a,b])$ and $a \in \mathbb{R}$, then

$$
\sum_{\pi_j \in \Pi} \left(\frac{|\Delta(af)(\pi_j)|}{|\pi_j|} \right)^p \omega(\pi_j) = |a| \sum_{\pi_j \in \Pi} \left(\frac{|\Delta f(\pi_j)|}{|\pi_j|} \right)^p \omega(\pi_j),
$$

and so

$$
\|af\|_{\mathrm{RBV}^p_\omega[a,b]} = |a| \|f\|_{\mathrm{RBV}^p_\omega[a,b]}.
$$

Thus $\| \cdot \|_{\mathrm{RBV}^p_\omega[a,b]}$ is a norm, and so $\mathrm{RBV}^p_\omega([a,b])$ is a normed linear space. □

Theorem 4.4. *The space $\mathrm{RBV}_\omega^p([a,b])$ with norm $\|\cdot\|_{\mathrm{RBV}_\omega^p([a,b])}$ is a Banach space.*

Proof. Let $\{f_j\}_{j\in\mathbb{N}}$ be a Cauchy sequence in $\mathrm{RBV}_\omega^p[a,b]$. Then given a partition Π of $[a,b]$, for any $\epsilon > 0$ there exits $N \in \mathbb{N}$ such that for $n, m \in \mathbb{N}$,

$$\sum_{\pi_j \in \Pi} \left(\frac{|\Delta(f_n - f_m)(\pi_j)|}{|\pi_j|} \right)^p \omega(\pi_j) \leq V_\omega^p(f_n - f_m, [a,b]) < \epsilon$$

and $|f_n(a) - f_m(a)| < \epsilon$. Therefore, the sequence $\{f_n(a)\}_{n\in\mathbb{N}}$ converges. Similarly, for $x > a$,

$$\begin{aligned} |f_n(x) - f_m(x)|^p &\leq 2^p |f_n(a) - f_m(a)|^p + 2^p \frac{|\Delta(f_n - f_m)([a,x])|^p}{(\sigma([a,x]))^{p-1}} (\sigma([a,x]))^p \\ &\leq 2^p |f_n(a) - f_m(a)|^p + 2^p |f_n(a) - f_m(a)|^p \\ &\quad + 2^p V_\omega^p(f_n - f_m, [a,b])(\sigma([a,x]))^p. \end{aligned}$$

Therefore, the sequence $\{f_n\}_{n\in\mathbb{N}}$ is uniformly Cauchy and so, by Proposition 4.4, it converges uniformly to a continuous function f. Consequently, we have for $n \geq N$,

$$\sum_{\pi_j \in \Pi} \left(\frac{|\Delta(f - f_m)(\pi_j)|}{|\pi_j|} \right)^p \omega(\pi_j) < \epsilon.$$

Since this is true for all partitions Π, we have

$$V_\omega^p(f - f_m) < \epsilon.$$

Therefore, we must have

$$\|f - f_m\|_{\mathrm{RBV}_\omega^p([a,b])} \to 0 \quad \text{as } m \to \infty.$$

By the triangle inequality,

$$\|f\|_{\mathrm{RBV}_\omega^p([a,b])} \leq \|f - f_m\|_{\mathrm{RBV}_\omega^p} + \|f_m\|_{\mathrm{RBV}_\omega^p} < \epsilon.$$

So $f \in \mathrm{RBV}_\omega^p$. Thus $\mathrm{RBV}_\omega^p([a,b])$ is a Banach space. □

Definition 4.5. A function $f : [a,b] \to \mathbb{R}$ is said to be Lipschitz continuous if there exists a constant $L \geq 0$ such that

$$|f(x) - f(y)| \leq L|x - y| \quad x, y \in [a,b].$$

The set of all Lipschitz functions is denoted by $\mathrm{Lip}([a,b])$.

Given $f : [a,b] \to \mathbb{R}$ and a partition $\mathcal{P} = \{a = t_0 < t_1 < \cdots < t_n = b\}$ of $[a,b]$, the variation of f over $[a,b]$ it is defined by

$$V(f,\mathcal{P})=\sum_{k=1}^{n}\left|f(t_k)-f(t_{k-1})\right|.$$

Also the total variation f over $[a,b]$ is defined by

$$V_a^b(f)=\sup_{\mathcal{P}} V(f,\mathcal{P}),$$

where the supremum is taken over all partitions $\mathcal{P}$ of $[a,b]$.

Definition 4.6. If $V_a^b(f)<\infty$, we say that f is a function of bounded variation.

Definition 4.7. A function f is absolutely continuous, denoted $f\in AC([a,b])$, if for every $\epsilon>0$, there exists $\delta>0$ such that given any finite collection $\{(a_j,b_j)\}_{j=1}^n$ in $[a,b]$,

$$\sum_{j=1}^{n}(b_j-a_j)<\delta \quad \text{implies} \quad \sum_{j=1}^{n}\left|f(b_j)-f(a_j)\right|<\epsilon.$$

Proposition 4.6. *Consider $\omega\in A_p$ and $1<p<\infty$. Then*

$$\mathrm{RBV}_\omega^p([a,b])\subset \mathrm{BV}([a,b]).$$

Proof. Let $\Pi=\{\pi_j\}_{j=1}^n$ be a finite partition of $[a,b]$ and $\sigma=\omega^{1-q}$. Then by Hölder's inequality, we have

$$\begin{aligned}
\sum_{\pi_j\in\Pi}|\Delta f(\pi_j)| &= \sum_{\pi_j\in\Pi}|\Delta f(\pi_j)|\frac{(\sigma(\pi_j))^{\frac{(p-1)}{p}}}{(\sigma(\pi_j))^{\frac{(p-1)}{p}}}\\
&\le \left[\sum_{\pi_j\in\Pi}\left(\frac{|\Delta f(\pi_j)|}{|\pi_j|}\right)^p\omega(\pi_j)\right]^{\frac{1}{p}}\left(\sum_{\pi_j\in\Pi}(\sigma(\pi_j))^{\frac{(p-1)q}{p}}\right)^{\frac{1}{q}}\\
&\le [V_\omega^p(f,[a,b])]^{\frac{1}{p}}\left(\sum_{\pi_j\in\Pi}\sigma(\pi_j)\right)^{\frac{1}{q}}\\
&\le (V_\omega^p(f,[a,b]))^{\frac{1}{p}}\sigma([a,b])<\infty.
\end{aligned}$$

Taking the supremum over all finite partitions Π of $[a,b]$, we get the desired result. □

Proposition 4.7. *Fix $1<p<\infty$ and assume $\frac{1}{p}+\frac{1}{q}=1$. Given a weight $\omega\in A_p$,*

$$\mathrm{Lip}([a,b])\subseteq \mathrm{RBV}_\omega^p([a,b]).$$

Proof. Let $\Pi = \{\pi_j\}$ be a finite partition of $[a, b]$ and $f \in \mathrm{Lip}([a, b])$. Then

$$\begin{aligned}\sum_{\pi_j\in\Pi}\left(\frac{|\Delta f(\pi_j)|}{|\pi_j|}\right)^p \omega(\pi_j) &= \sum_{\pi_j\in\Pi}\left(\frac{|f(x_j)-f(x_{j-1})|}{|x_j - x_{j-1}|}\right)^p \omega(\pi_j)\\ &= \sum_{\pi_j\in\Pi}\left(\frac{L|x_j - x_{j-1}|}{|x_j - x_{j-1}|}\right)^p \omega(\pi_j)\\ &= L^p \sum_{\pi_j\in\Pi}\omega(\pi_j)\\ &\le L^p\omega([a,b]) < \infty.\end{aligned}$$

Taking the supremum over all finite partitions Π of $[a, b]$, we get the desired result. □

Finally, we shall show that the spaces $\mathrm{RBV}^p_\omega([a, b])$ are naturally embedded in one another.

Proposition 4.8. *Given a weight $\omega \in A_p$, let $1 < p \le q < \infty$. Then we have*

$$\mathrm{RBV}^q_\omega([a,b]) \subseteq \mathrm{RBV}^p_\omega([a,b]).$$

Proof. The case $p = 1$ follows from Proposition 4.6, so we may assume that $1 < p < q < \infty$. Let

$$r = \frac{q}{p}, \quad s = \frac{q}{q-p}$$

and observe that $\frac{1}{r} + \frac{1}{s} = 1$. Therefore, by Hölder's inequality, we have

$$\begin{aligned}\sum_{\pi_j\in\Pi}\frac{|\Delta f(\pi_j)|^p}{(\sigma(\pi_j))^{p-1}} &= \sum_{\pi_j\in\Pi}\frac{|\Delta f(\pi_j)|^p}{(\sigma(\pi_j))^{p-1+\frac{1}{s}}}(\sigma(\pi_j))^{\frac{1}{s}}\\ &\le \left(\sum_{\pi_j\in\Pi}\frac{|\Delta f(\pi_j)|^{pr}}{(\sigma(\pi_j))^{(p-1+\frac{1}{s})r}}\right)^{\frac{1}{r}}\left(\sum_{\pi_j\in\Pi}\sigma(\pi_j)\right)^{\frac{1}{s}}\\ &\le \left(\sum_{\pi_j\in\Pi}\left(\frac{|\Delta f(\pi_j)|^{pr}}{|\pi_j|}\right)^q \omega(\pi_j)\right)^{\frac{1}{r}}(\sigma([a,b]))^{\frac{1}{s}} < \infty.\end{aligned}$$

Hence

$$V^p_\omega(f, [a,b]) < \infty.$$

□

Proposition 4.9. *Given a weight $\omega \in A_p$ and $1 < p < \infty$,*

$$\mathrm{RBV}^p_\omega([a,b]) \subset \mathrm{AC}([a,b]).$$

Proof. Since $\sigma = \omega^{1-q} \in L_1([a,b])$, by the continuity of the integral, for any $\epsilon > 0$ there exists $\delta > 0$ such that if $|E| < \delta$, then $\sigma(E) < \epsilon$. Fix any collection $\{(a_j, b_j)\}_{j=1}^{n}$ of disjoint opens balls in $[a,b]$ such that

$$\sum_{j=1}^{n}(b_j - a_j) < \delta.$$

Let

$$E = \bigcup_{j=1}^{n}[a_j, b_j].$$

Then $\sigma(E) < \epsilon$.

If we let $\frac{1}{p} + \frac{1}{q} = 1$, then by Hölder's inequality in sequence spaces,

$$\begin{aligned}\sum_{j=1}^{n}|f(b_j) - f(a_j)| &= \sum_{j=1}^{n}\frac{|f(b_j) - f(a_j)|}{(\sigma([a_j,b_j]))^{\frac{1}{q}}}(\sigma([a_j,b_j]))^{\frac{1}{q}}\\ &\le \left(\sum_{j=1}^{n}\left(\frac{|f(b_j) - f(a_j)|}{|[a_j,b_j]|}\right)^{p}\omega([a_j,b_j])\right)^{\frac{1}{p}}\left(\sum_{j=1}^{n}\sigma([a_j,b_j])\right)^{\frac{1}{q}}\\ &\le (V_{\omega}^{p}(f,[a,b]))^{\frac{1}{p}}\epsilon^{\frac{1}{q}}.\end{aligned}$$

Since $\epsilon > 0$ is arbitrary, it follows that $f \in \mathrm{AC}([a,b])$. □

Finally, we can now state and prove the weighted version of the Riesz theorem.

Theorem 4.5. *Let $1 < p < \infty$ and suppose $\frac{1}{p} + \frac{1}{q} = 1$. Given a weight $\omega \in A_p$, we have $f \in \mathrm{RBV}_{\omega}^{p}([a,b])$ if and only if $f \in \mathrm{AC}([a,b])$ and $f' \in L_p(\omega)$. Moreover,*

$$\frac{1}{[\omega]_{A_p}}V_{\omega}^{p}(f,[a,b]) \le \|f'\|_{L_p(\omega)}^{p} \le V_{\omega}^{p}(f,[a,b]).$$

Proof. Suppose first that $f \in \mathrm{AC}([a,b])$ and $f' \in L_p(\omega)$. Fix a finite partition $\Pi = \{\pi_j\}$ of $[a,b]$. Using the fact that $f \in \mathrm{AC}([a,b])$, we can estimate as follows:

$$\begin{aligned}\left(\frac{|\Delta f(\pi_j)|}{|\pi_j|}\right)^{p}\omega(\pi_j) &\le [\omega]_{A_p}\left|\int_{\pi_j}f'(t)\,dt\right|^{p}(\sigma(\pi_j))^{1-p}\\ &\le [\omega]_{p}\left(\int_{\pi_j}|f'(t)|\omega^{\frac{1}{p}}\omega^{-\frac{1}{p}}\,dt\right)^{p}(\sigma(\pi_j))^{1-p}\\ &\le [\omega]_{A_p}\left[\left(\int_{\pi_j}|f'(t)|\omega\,dt\right)^{\frac{1}{p}}(\sigma(t)\,dt)^{\frac{1}{q}}\right]^{p}(\sigma(\pi_j))^{1-p}\end{aligned}$$

$$= [\omega]_{A_p}\Big(\int_{\pi_j} |f'(t)|\omega\,dt\Big)(\sigma(\pi_j))^{\frac{p}{q}}(\sigma(\pi_j))^{1-p}$$

$$= [\omega]_{A_p}\int_{\pi_j} |f'(t)|^p\omega(t)\,dt.$$

Summing over the subintervals, we obtain

$$\sum_{\pi_j\in\Pi}\Big(\frac{|\Delta f(\pi_j)|}{|\pi_j|}\Big)^p\omega(\pi_j) \le [\omega]_{A_p}\int_{[a,b]} |f'(t)|^p\omega(t)\,dt,$$

which implies that

$$V^p_\omega(f,[a,b]) \le [\omega]_{A_p}\|f'\|^p_{L_p(\omega)}.$$

So

$$\frac{1}{[\omega]_{A_p}}V^p_\omega(f,[a,b]) \le \|f'\|^p_{L_p(\omega)}.$$

To prove the reverse inequality, let $f \in \mathrm{RBV}^p_\omega([a,b])$. Define the regular partition $\Pi_N = \{[t_{k,N}, t_{k+1,N}]\}_{k=0}^{N-1}$ where $t_{k,N} = a + \frac{k}{N}(b-a)$.

For brevity, let $\Pi_{j,N} = [t_{j-1,N}, t_{j,N}]$. By Proposition 4.9, $f \in \mathrm{AC}([a,b])$, so f' exists in $L_1([a,b])$. Therefore, we may define

$$g_N(t) = \sum_{\pi_{j,N}\in\Pi_N} \fint_{\pi_{j,N}} f'(x)\chi_{\pi_{j,N}}(t)\,dx.$$

Then, by the Lebesgue differentiation theorem, we obtain

$$\lim_{N\to\infty} g_N(t) = f'(t),$$

for almost every $t \in [a,b]$.

Consequently, by Fatou's lemma applied with respect to the measure $\omega\,dx$ and using the fact that $\omega \in A_p$, we estimate as follows:

$$\begin{aligned}
\int_{[a,b]} |f'(t)|^p\omega(t)\,dt &\le \liminf_{N\to\infty}\int_{[a,b]} |g_N(t)|^p\omega(t)\,dt\\
&= \liminf_{N\to\infty}\sum_{\pi_{j,N}\in\Pi_N}\int_{\pi_{j,N}} |g_N(t)|^p\omega(t)\,dt\\
&\le \liminf_{N\to\infty}\sum_{\pi_{j,N}\in\Pi_N}\Big(\frac{|\Delta f(\pi_{j,N})|}{|\pi_{j,N}|}\Big)^p\int_{\pi_{j,N}}\omega(t)\,dt\\
&= \liminf_{N\to\infty}\sum_{\pi_{j,N}\in\Pi_N}\Big(\frac{|\Delta f(\pi_{j,N})|}{|\pi_{j,N}|}\Big)^p\omega(\pi_{j,N})\\
&\le V^p_\omega(f,[a,b]) < \infty.
\end{aligned}$$

So

$$\|f'\|^p_{L_p(\omega)} \le V^p_\omega(f, [a, b]).$$

Finally, we have

$$\frac{1}{[\omega]_{A_p}} V^p_\omega(f, [a, b]) \le \|f'\|^p_{L_p(\omega)} \le V^p_\omega(f, [a, b]),$$

as needed. □

4.3 The Gehring lemma

Assume that $f : \mathbb{R}^n \to [-\infty, +\infty]$ is a Lebesgue measurable function. According to Hölder's or Jensen's inequality,

$$\fint_Q |f(x)|\,dx \le \left(\fint_Q |f(x)|^p\,dx\right)^{1/p} \left(\fint_Q \chi_Q(x)\,dx\right)^{1/p} = \left(\fint_Q |f(x)|^p\,dx\right)^{1/p}$$

for any cube $Q \subset \mathbb{R}^n$. Next, we consider uniform inequalities over cubes in the reverse direction.

Theorem 4.6 (The Gehring lemma (see [39])). *Assume that $f \in L^1_{\mathrm{loc}}(\mathbb{R}^n)$. Let $1 < p < +\infty$ and assume that there exists a constant $c > 0$ such that*

$$\left(\fint_Q |f(x)|^p\,dx\right)^{1/p} \le c \fint_Q |f(x)|\,dx \tag{4.23}$$

for every cube $Q \subset \mathbb{R}^n$. Then there exist $q > p$ and constant $c' = c'(n, p, q, c) > 0$ such that

$$\left(\fint_Q |f(x)|^q\,dx\right)^{1/q} \le c' \fint_Q |f(x)|\,dx \tag{4.24}$$

for every cube $Q \subset \mathbb{R}^n$.

Proof. Step 1. Considering any cube $Q \subset \mathbb{R}^n$ and denoting

$$E_t = \{x \in Q : |f(x)| > t\}, \quad t > 0,$$

we claim that

$$\int_{E_{2t}} |f(x)|^p\,dx \le c_1 t^{p-1} \int_{E_t} |f(x)|\,dx \quad \text{wherever } t \ge |f|_Q, \tag{4.25}$$

with $c_1 = (2^{n+1}c)^p$. The point here is that the information over cubes is retained in the information over the distribution sets.

Denote $s = 2t$. By Calderón–Zygmund decomposition in Q at level $s = 2t \geq 2\|f\|_Q \geq |f|_Q$, there are countably or finitely many dyadic subcubes Q_i, $i = 1, 2, \dots$, of Q such that the interiors of Q_i, $i = 1, 2, \dots$, are pairwise disjoint,

$$s < \fint_{Q_i} |f(x)|\, dx \leq 2^n s, \quad i = 1, 2, \dots,$$

and $|f(x)| \leq s$ for almost every $x \in Q_0 \setminus \bigcup_{i=1}^{+\infty} Q_i$. Thus the Calderón–Zygmund cubes cover the set $E_s = \{x \in Q : |f(x)| > s\}$ up to a set of measure zero and thus, by (4.23), we have

$$\begin{aligned}
\int_{E_s} |f(x)|^p\, dx &\leq \sum_{i=1}^{+\infty} \int_{Q_i} |f(x)|^p\, dx \\
&\leq \sum_{i=1}^{+\infty} \left(c \fint_{Q_i} |f(x)|\, dx \right)^p m(Q_i) \\
&\leq \sum_{i=1}^{+\infty} (c2^n s)^p m(Q_i) \\
&= (c2^n s)^p \sum_{i=1}^{+\infty} m(Q_i), \quad i = 1, 2, \dots.
\end{aligned} \tag{4.26}$$

On the other hand,

$$\begin{aligned}
m(Q_i) \leq \frac{1}{s} \int_{Q_i} |f(x)|\, dx &= \frac{1}{s} \int_{Q_i \cap E_t} |f(x)|\, dx + \frac{1}{s} \int_{Q_i \setminus E_t} |f(x)|\, dx \\
&\leq \frac{1}{s} \int_{Q_i \cap E_t} |f(x)|\, dx + \frac{t}{s} m(Q_i), \quad i = 1, 2, \dots,
\end{aligned}$$

where we estimated

$$\int_{Q_i \setminus E_t} |f(x)|\, dx \leq \int_{Q_i \setminus E_t} t\, dx = t\, m(Q_i \setminus E_t) \leq t\, m(Q_i), \quad i = 1, 2, \dots.$$

Since $\frac{t}{s} = \frac{t}{2t} = \frac{1}{2}$, we may absorb $\frac{1}{2} m(Q_i)$ on the right-hand side and obtain

$$\frac{1}{2} m(Q_i) = \frac{1}{s} \int_{Q_i \cap E_t} |f(x)|\, dx, \quad i = 1, 2, \dots,$$

which implies

$$m(Q_i) \leq \frac{2}{s} \int_{Q_i \cap E_t} |f(x)|\, dx = \frac{1}{t} \int_{Q_i \cap E_t} |f(x)|\, dx, \quad i = 1, 2, \dots.$$

By inserting this into (4.26), we obtain

$$\begin{aligned}\int_{E_t} |f(x)|^p\,dx &\le (2^m ks)^p \sum_{i=1}^{+\infty} m(Q_i)\\ &\le (2^m cs)^p \sum_{i=1}^{+\infty} \frac{1}{t} \int_{Q_i\cap E_t} |f(x)|\,dx\\ &= \frac{(2^m cs)^p}{t} \sum_{i=1}^{+\infty} \int_{Q_i\cap E_t} |f(x)|\,dx\\ &\le \frac{(2^m cs)^p}{t} \int_{E_t} |f(x)|\,dx\\ &= (2^m c)^p t^{p-1} \int_{E_t} |f(x)|\,dx, \quad i = 1, 2, \ldots .\end{aligned}$$

Here we used the fact that the Calderón–Zygmund cubes Q_i, $i = 1, 2, \ldots$ are pairwise disjoint and $s = 2t$.

Step 2. We claim that

$$\int_{E_t} |f(x)|^p\,dx \le c_2 t^{p-1} \int_{E_t} |f(x)|\,dx \quad \text{wherever } t \ge |f|_Q, \tag{4.27}$$

with $c_2 = (2^{p-1} + (2^{n-1}c))^p$. Observe that $E_s \subseteq E_t$ with $s = 2t$ and thus

$$\int_{E_t} |f(x)|^p\,dx = \int_{E_t\setminus E_s} |f(x)|^p\,dx + \int_{E_s} |f(x)|^p\,dx.$$

For the second term on the right-hand side, we obtained is step 1 that

$$\int_{E_s} |f(x)|^p\,dx \le (2^{n+1}c)^p t^{p-1} \int_{E_t} |f(x)|^p\,dx.$$

For the first term on the right-hand side, we obtain

$$\begin{aligned}\int_{E_t\setminus E_s} |f(x)|^p\,dx = \int_{E_t\setminus E_s} |f(x)|^{p-1}|f(x)|\,dx &\le s^{p-1} \int_{E_t\setminus E_s} |f(x)|\,dx\\ &\le (2t)^{p-1} \int_{E_t} |f(x)|\,dx\\ &= 2^{p-1}t^{p-1} \int_{E_t} |f(x)|\,dx.\end{aligned}$$

Here we also used the fact that $t < |f(x)| \le s$ for every $x \in E_t \setminus E_s$. By combining these estimates, we conclude that

$$\begin{aligned}\int_{E_t} |f(x)|^p\,dx &= \int_{E_t\setminus E_s} |f(x)|^p\,dx + \int_{E_t} |f(x)|^p\,dx\\ &\le 2^{p-1}t^{p-1} \int_{E_t} |f(x)|\,dx + (2^{n+1}c)^p t^{p-1} \int_{E_t} |f(x)|\,dx\\ &= (2^{p-1} + (2^{n+1}c)^p)^p t^{p-1} \int_{E_t} |f(x)|\,dx.\end{aligned}$$

Step 3. Denote $t_0 = |f|_Q$ and assume that $q > p$. Then a similar argument as in the proof of the Cavalieri's principle with $t = t_0$ gives

$$\begin{aligned}
\int_{E_{t_0}} |f(x)|^q \, dx &= \int_{E_{t_0}} |f(x)|^p |f(x)|^{q-p} \, dx \\
&= \int_{E_{t_0}} |f(x)|^p \left(t_0^{q-p} + (q-p) \int_{t_0}^{|f(x)|} t^{q-p-1} \, dt \right) dx \\
&= t_0^{q-p} \int_{E_{t_0}} |f(x)|^p \, dx + (q-p) \int_{E_{t_0}} |f(x)|^p \int_{E_{t_0}}^{|f(x)|} t^{q-p-1} \, dt \, dx \\
&\le c_2 \, t_0^{q-1} \int_{E_{t_0}} |f(x)|^p \, dx + (q-p) \int_{E_{t_0}} |f(x)|^p \int_{E_{t_0}}^{|f(x)|} t^{q-p-1} \, dt \, dx.
\end{aligned}$$

By Fubini's theorem,

$$\begin{aligned}
\int_{E_{t_0}} |f(x)|^p \int_{E_{t_0}}^{|f(x)|} t^{q-p-1} \, dt \, dx &= \int_{\mathbb{R}^n} \mathcal{X}_{E_{t_0}}(x) |f(x)|^p \int_0^{+\infty} \mathcal{X}_{[t_0, |f(x)|]}(t) \, t^{q-p-1} \, dt \, dx \\
&= \int_{\mathbb{R}^n} \int_0^{+\infty} \mathcal{X}_{E_{t_0}}(x) |f(x)|^p \mathcal{X}_{[t_0, |f(x)|]}(t) \, t^{q-p-1} \, dt \, dx \\
&= \int_0^{+\infty} \int_{\mathbb{R}^n} \mathcal{X}_{E_{t_0}}(x) |f(x)|^p \mathcal{X}_{[t_0, +\infty]}(t) \, t^{q-p-1} \, dx \, dt \\
&= \int_{t_0}^{+\infty} t^{q-p-1} \int_{E_t} |f(x)|^p \, dx \, dt.
\end{aligned}$$

Here we used the fact that

$$\mathcal{X}_{E_{t_0}}(x) \mathcal{X}_{[t_0, |f(x)|]}(t) = \mathcal{X}_{[t_0, +\infty]}(t) \mathcal{X}_{E_t}(x).$$

By (4.27) with $t \ge t_0 = |f|_Q$, we have

$$t^{q-p} \int_{E_t} |f(x)|^p \, dx \le t^{q-p} \int_{E_t} |f(x)| \, dx = c_2 \, t^{q-1} \int_{E_t} |f(x)| \, dx$$

and

$$\begin{aligned}
\int_{t_0}^{+\infty} t^{q-p-1} \int_{E_t} |f(x)|^p \, dx \, dt &\le c_2 \int_{t_0}^{+\infty} t^{q-p-1} t^{p-1} \int_{E_t} |f(x)|^p \, dx \, dt \\
&= c_2 \int_{t_0}^{+\infty} t^{q-2} \int_{E_t} |f(x)|^p \, dx \, dt \\
&= c_2 \int_{t_0}^{+\infty} \int_{\mathbb{R}^n} \mathcal{X}_{E_t}(x) |f(x)| \mathcal{X}_{[t_0, +\infty)}(t) \, t^{q-2} \, dx \, dt \\
&= c_2 \int_{t_0}^{+\infty} \int_{\mathbb{R}^n} \mathcal{X}_{E_{t_0}}(x) |f(x)| \mathcal{X}_{[t_0, |f(x)|]}(t) \, t^{q-2} \, dx \, dt
\end{aligned}$$

$$
\begin{aligned}
&= c_2 \int_{\mathbb{R}^n} \int_{t_0}^{+\infty} \mathcal{X}_{E_{t_0}}(x)|f(x)|\mathcal{X}_{[t_0,|f(x)|]}(t)\, t^{q-2}\, dt\, dx \\
&= c_2 \int_{\mathbb{R}^n} \mathcal{X}_{E_{t_0}}(x)|f(x)| \int_{t_0}^{+\infty} \mathcal{X}_{[t_0,|f(x)|]}(t)\, t^{q-2}\, dt\, dx \\
&= c_2 \int_{E_t} |f(x)| \int_{t_0}^{|f(x)|} t^{q-2}\, dt\, dx \\
&= \frac{c_2}{q-1} \int_{E_{t_0}} (|f(x)||f(x)|^{q-1} - t_0^{q-1}|f(x)|)\, dx \\
&= \frac{c_2}{q-1}\left(\int_{E_{t_0}} |f(x)|^q\, dx - t_0^{q-1} \int_{E_{t_0}} |f(x)|\, dx \right).
\end{aligned}
$$

Thus we conclude that

$$
\begin{aligned}
\int_{t_0} |f(x)|^q\, dx &\leq c_2\, t_0^{q-1} \int_{E_{t_0}} |f(x)|\, dx + (q-p) \int_{E_t} |f(x)|^p \int_{t_0}^{|f(x)|} t^{q-p-1} t^{p-1}\, dt\, dx \\
&\leq c_2\, t_0^{q-1} \int_{E_{t_0}} |f(x)|\, dx + (q-p) \int_{t_0}^{+\infty} t^{q-p-1} \int_{E_t} |f(x)|^p\, dx\, dt \\
&\leq c_2\, t_0^{q-1} \int_{E_{t_0}} |f(x)|\, dx + c_2 \frac{(q-p)}{q-1}\left(\int_{E_{t_0}} |f(x)|^q\, dx - t_0^{q-1} \int_{E_{t_0}} |f(x)|\, dx \right) \\
&\leq c_2 \frac{(q-p)}{q-1} \int_{E_{t_0}} |f(x)|\, dx + (c_2 - 1)\, t_0^{q-1} \int_{E_{t_0}} |f(x)|\, dx.
\end{aligned}
$$

If $\int_{E_{t_0}} |f(x)|^q\, dx < +\infty$, we may transfer the first term on the right-hand side to the left-hand side and obtain

$$
\left(1 - c_2 \frac{(q-p)}{q-1}\right) \int_{E_{t_0}} |f(x)|^q\, dx \leq c_2\, t^{q-1} \int_{E_{t_0}} |f(x)|\, dx. \tag{4.28}
$$

Step 4. Note that at this point we do not know if $\int_{E_{t_0}} |f(x)|^q\, dx < +\infty$, since this is essentially what we want to prove in Gehring's lemma. To conclude this, we consider truncations

$$
f_k = f\, \mathcal{X}_{\{|f| \leq k\}}, \quad k = 1, 2, \dots
$$

and note that, due to (4.27), functions f_k, $k = 1, 2, \dots,$ satisfy

$$
\int_{\{x \in Q : |f_k(x)| > t\}} |f(x)|^p\, dx \leq c_2\, t^{p-1} \int_{\{x \in Q : |f_k(x)| > t\}} |f(x)|^p\, dx, \quad t \leq t_0 = |f|_Q,
$$

and

$$
\int_{E_{t_0}} |f(x)|^{q-p} |f(x)|^p\, dx \leq k^{q-p} \int_{E_{t_0}} |f(x)|^p\, dx < +\infty, \quad k = 1, 2, \dots.
$$

By the same argument as in step 3, together with (4.28), we have

$$\left(1 - c_2 \frac{(q-p)}{q-1}\right) \int_{E_{t_0}} |f_k(x)|^{q-p} |f(x)|^p \, dx \le c_2 \, t_0^{q-1} \int_{E_{t_0}} |f_k(x)| \, dx \, , \quad k = 1, 2, \ldots ,$$

and by the Lebesgue monotone convergence theorem,

$$\begin{aligned} \left(1 - c_2 \frac{(q-p)}{q-1}\right) \int_{E_{t_0}} |f(x)|^q \, dx &= \left(1 - c_2 \frac{(q-p)}{q-1}\right) \lim_{k \to +\infty} \int_{E_{t_0}} |f_k(x)|^{q-1} |f(x)| \, dx \\ &\le c_2 \, t_0^{q-1} \int_{E_{t_0}} |f(x)|^q \, dx < +\infty \, . \end{aligned}$$

By choosing $q > p$ such that

$$1 - c_2 \frac{(q-p)}{q-1} = \frac{1}{2},$$

we obtain

$$\int_{E_{t_0}} |f(x)|^q \, dx \le 2 \, c_2 \, t_0^{q-1} \int_{E_{t_0}} |f(x)| \, dx \, .$$

In particular, this implies $\int_{E_{t_0}} |f(x)|^q \, dx < +\infty$.

Step 5. Finally, we show that (4.24) holds true. Indeed,

$$\begin{aligned} \int_Q |f(x)|^q \, dx &= \int_{Q \setminus E_{t_0}} |f(x)|^q \, dx + \int_{E_{t_0}} |f(x)|^q \, dx \\ &\le \int_{Q \setminus E_{t_0}} |f(x)|^{q-1} |f(x)| \, dx + 2 \, c_2 \, t_0^{q-1} \int_{E_{t_0}} |f(x)| \, dx \\ &\le t_0^{q-1} \int_{Q \setminus E_{t_0}} f(x)| \, dx + 2 \, c_2 \, t_0^{q-1} \int_{E_{t_0}} |f(x)| \, dx \\ &\le 2 \, c_2 \, t_0^{q-1} \int_Q |f(x)| \, dx \, . \end{aligned}$$

Here we used the fact that $|f(x)|^{q-1} \le t_0^{q-1}$ for every $x \in Q \setminus E_{t_0}$.

By inserting $t_0 = |f|_Q = \int_Q |f(x)| \, dx$, we arrive at

$$\int_Q |f(x)|^q \, dx \le c \left(\int_Q |f(x)| \, dx \right)^{q-1} \int_Q |f(x)| \, dx,$$

from which (4.24) follows. □

4.4 Reverse Hölder's inequality

The next theorem shows that every A_p weight satisfies a reverse Hölder's inequality.

Theorem 4.7. *Let $1 < p < +\infty$ and assume that $\omega \in A_p$. Then there are $\delta > 0$ and $0 < c < +\infty$ such that*

$$\left(⨍_Q (\omega(x))^{1+\delta}\, dx\right)^{\frac{1}{1+\delta}} \le c ⨍_Q \omega(x)\, dx$$

for every cube $Q \subset \mathbb{R}^n$.

Proof. By Theorem 4.1, we have $A_p \subset A_q$ if $1 \le p < q$. Thus we may assume that $p > 2$. By Hölder's inequality, for every $f > 0$ and every $Q \subset \mathbb{R}^n$, we have

$$\begin{aligned} m(Q) &= \int_Q dx,\\ 1 &= \frac{1}{m(Q)} \int_Q \sqrt{f} \frac{1}{\sqrt{f}}\, dx\\ &= \int_Q \frac{\sqrt{f}}{\sqrt{m(Q)}} \frac{1}{\sqrt{m(Q)}} \frac{1}{\sqrt{f}}\, dx\\ &\le \left(\frac{1}{m(Q)} \int_Q f\, dx\right)^{1/2} \left(\frac{1}{m(Q)} \int_Q \frac{1}{f}\, dx\right)^{1/2}. \end{aligned}$$

Thus

$$1 \le \left(⨍_Q f\, dx\right)\left(⨍_Q \frac{1}{f}\, dx\right).$$

Inserting $f = \omega^{1/(p-1)}$ gives

$$1 \le ⨍_Q \left(\frac{1}{\omega(x)}\right)^{\frac{1}{p-1}} dx ⨍_Q (\omega(x))^{\frac{1}{p-1}}\, dx.$$

Since $\omega \in A_p$, we obtain

$$\begin{aligned} ⨍_Q \omega(x)\, dx \left(⨍_Q (\omega(x))^{\frac{1}{1-p}}\, dx\right)^{p-1} &\le c = c\, 1^{p-1}\\ &\le c\left(⨍_Q \left(\frac{1}{\omega(x)}\right)^{\frac{1}{p-1}} dx\right)^{p-1} \left(⨍_Q (\omega(x))^{\frac{1}{p-1}}\, dx\right)^{p-1}. \end{aligned}$$

Dividing both sides by $0 < (⨍_Q (\omega(x))^{\frac{1}{p-1}}\, dx)^{p-1} < +\infty$, we obtain

$$\fint_Q \omega(x)\,dx \le c\left(\fint_Q (\omega(x))^{\frac{1}{p-1}}\,dx\right)^{p-1},$$

and thus

$$\left(\left(\fint_Q (\omega(x))^{\frac{1}{p-1}}\right)^{p-1} dx\right)^{\frac{1}{p-1}} \le c \fint_Q (\omega(x))^{\frac{1}{p-1}}\,dx$$

for every cube $Q \subset \mathbb{R}^n$. Since $p - 1 > 1$, this shows that $(\omega(x))^{\frac{1}{p-1}}$ satisfies a reverse Hölder's inequality. Then, there exist $q > p - 1$ and $c' < +\infty$ such that

$$\begin{aligned}\left(\fint_Q (\omega(x))^{\frac{q}{p-1}}\,dx\right)^{\frac{1}{q}} &\le c' \fint_Q (\omega(x))^{\frac{1}{p-1}}\,dx \\ &\le c'(m(Q))^{\frac{p-2}{p-1}}\left(\fint_Q \omega(x)\,dx\right)^{\frac{1}{p-1}} \\ &= c\left(\fint_Q \omega(x)\,dx\right)^{\frac{1}{p-1}},\end{aligned}$$

where $c = c'(m(Q))^{\frac{p-2}{p-1}}$.

Hence

$$\left(\fint_Q (\omega(x))^{\frac{q}{p-1}}\,dx\right)^{\frac{1}{q}} \le c\left(\fint_Q \omega(x)\,dx\right)^{\frac{1}{p-1}}.$$

Thus we arrive at

$$\left(\fint_Q (\omega(x))^{\frac{q}{p-1}}\,dx\right)^{\frac{1}{q}} \le c \fint_Q \omega(x)\,dx,$$

for every $Q \subset \mathbb{R}^n$. The claim now follows by choosing $\delta > 0$ such that $1 + \delta = \frac{q}{p-1}$. □

4.5 The Rubio de Francia algorithm

The Rubio de Francia algorithm stands as a cornerstone of modern analysis, offering a powerful and systematic framework for studying the structure of weighted Lorentz spaces. Developed through the foundational work of José L. Rubio de Francia, this innovative computational tool has transformed our understanding of these function spaces by providing a methodical approach to constructing atomic decompositions adapted to their unique properties.

Beyond its theoretical elegance, the algorithm has far-reaching implications, influencing fields such as harmonic analysis, signal processing, and approximation theory.

Its versatility continues to inspire new insights and applications, reinforcing its significance in contemporary research. For further exploration, the reader may consult references [28, 29, 33, 62].

Theorem 4.8. *Fix $1 < p < \infty$ and $\omega \in A_p$. For any nonnegative function $h \in L_p(\omega)$, define*

$$Rh(x) = \sum_{k=0}^{\infty} \frac{M^k h(x)}{2^k \|M\|^k},$$

where for $k > 0$, $M^k h = M \circ M \circ \cdots \circ Mh$ denotes the kth iterate of the maximal operator and $M^0 h = h$. Then:

(1) $h(x) \le Rh(x)$,

(2) $\|Rh\|_{L_p(\omega)} \le 2\|h\|_{L_p(\omega)}$,

(3) $Rh \in A_1$ *and* $[Rh]_1 \le 2\|M\|$.

Proof. Observe that

$$h(x) = M^0 h(x) \le \sum_{k=0}^{\infty} \frac{M^k h(x)}{2^k \|M\|^k},$$

that is,

$$h(x) \le Rh(x).$$

To prove (2), we use Minkowski inequality and the fact that M^k is a linear operator since M is a linear operator. Indeed,

$$\begin{aligned}
\|Rh\|_{L_p(\omega)} &= \left\| \sum_{k=0}^{\infty} \frac{M^k h}{2^k \|M\|^k} \right\|_{L_p(\omega)} \\
&\le \sum_{k=0}^{\infty} \left\| \frac{M^k h}{2^k \|M\|^k} \right\|_{L_p(\omega)} \\
&\le \sum_{k=0}^{\infty} \frac{1}{2^k \|M\|^k} \|M^k h\|_{L_p(\omega)} \\
&\le \sum_{k=0}^{\infty} \frac{1}{2^k \|M\|^k} \|M\|^k \|h\|_{L_p(\omega)} \\
&= \sum_{k=0}^{\infty} \frac{\|h\|_{L_p(\omega)}}{2^k} \\
&= \left(\sum_{k=0}^{\infty} \frac{1}{2^k} \right) \|h\|_{p(\omega)} \\
&= 2\|h\|_{L_p(\omega)}.
\end{aligned}$$

Then $\|Rh\|_{L_p(\omega)} \le 2\|h\|_{L_p(\omega)}$. Now we can write

$$\begin{aligned} M(Rh)(x) &= M\left(\sum_{k=0}^{\infty} \frac{M^k h(x)}{2^k \|M\|^k}\right) \\ &\le \sum_{k=0}^{\infty} M\left(\frac{M^{k+1}h(x)}{2^k\|M\|^k}\right) \\ &= \sum_{k=0}^{\infty} \frac{M^{k+1}h(x)}{2^k\|M\|} \\ &= \sum_{k=0}^{\infty} 2\|M\| \frac{M^{k+1}h(x)}{2^{k+1}\|M\|^{k+1}} \\ &= 2\|M\|Rh(x), \end{aligned}$$

and so

$$\begin{aligned} & M(Rh)(x) \le 2\|M\|Rh(x) \\ \Longrightarrow\ & M(Rh)(x) \le 2\|M\| \sup_{x\in Q} Rh(x) \\ \Longrightarrow\ & M(Rh)(x)\left[\sup_{x\in Q} Rh(x)\right]^{-1} \le 2\|M\| \\ \Longrightarrow\ & [Rh]_1 \le \|M\|, \end{aligned}$$

which implies $Rh \in A_1$. □

As a consequence of Theorem 4.8, we can define the operator $M'f = M(f\omega)/\omega$. Since $\omega^{1-q} \in A_q$, M is bounded on $L_q(\omega^{1-q})$, and so M' is bounded on $L_{q(\omega)}$. Therefore, we can define another iteration algorithm:

$$R'h(x) = \sum_{k=0}^{\infty} \frac{(M')^k h(x)}{2^k\|M'\|^k}.$$

(Again $(M')^0 h = h$). Arguing exactly as in the proof of Theorem 4.8, we have that:
1. For all x, $|h(x)| \le R'h(x)$,
2. $\|R'h\|_{L_q(\omega)} \le 2\|h\|_{L_q(\omega)}$,
3. $M'(R'h)(x) \le 2\|M'\|$, and so $R'h\omega \in A_1$ with $[R'h\omega]_{A_1} \le 2\|M'\|$.

The proof of the following theorem is based on the argument given in [31].

Theorem 4.9. *Given an operator T, suppose that for some p_0, $1 \le p_0 < \infty$, and every $\omega \in A_{p_0}$, there exists a constant C depending on $[\omega]_{A_{p_0}}$ such that*

$$\int_{\mathbb{R}^n} |Tf(x)|^{p_0} \omega(x)\,dx \le C \int_{\mathbb{R}^n} |f(x)|^{p_0} \omega(x)\,dx.$$

Then, for every p, $1 < p < \infty$, and every $\omega \in A_p$, there exists a constant depending on $[\omega]_{A_p}$ such that

$$\int_{\mathbb{R}^n} |Tf(x)|^{p} \omega(x)\,dx \le C \int_{\mathbb{R}^n} |f(x)|^{p} \omega(x)\,dx.$$

Proof. By duality, there exists a nonnegative function $h \in L_q(\omega)$, $\|h\|_{L_q(\omega)} = 1$ such that

$$\begin{aligned}\|Tf\|_{L_p(\omega)} &= \int_{\mathbb{R}^n} |Tf(x)| h(x)\omega(x)\,dx \\ &\le \int_{\mathbb{R}^n} |Tf(x)| (Rf(x))^{-\frac{1}{q_0}} (Rf(x))^{\frac{1}{q_0}} R'h(x)\omega(x)\,dx,\end{aligned}$$

where we have used that $h \le R'h$ and, if $p_0 = 1$, we let $\frac{1}{q_0} = 0$. Since $Rf, R'h \in A_1$, by the reverse factorization Proposition 4.2(9), we have $(Rf)^{1-p_0} R'\omega \in A_{p_0}$. Therefore, by Hölder's inequality with respect to the measure $R'h\omega$ (if $p_0 > 1$), by our hypothesis and since $|f| \le Rf$,

$$\begin{aligned}\|Tf\|_{L_p(\omega)} &\le \left(\int_{\mathbb{R}^n} |f(x)|^{p_0} (Rf(x))^{1-p_0} R'h(x)\omega(x)\,dx\right)^{\frac{1}{p_0}} \left(\int_{\mathbb{R}^n} R(f(x)) R'h(x)\omega(x)\,dx\right)^{\frac{1}{q_0}} \\ &\le C\left(\int_{\mathbb{R}^n} |f(x)|^{p_0} (Rf(x))^{1-p_0} R'h(x)\omega(x)\,dx\right)^{\frac{1}{p_0}} \left(\int_{\mathbb{R}^n} Rf(x) R'h(x)\omega(x)\,dx\right)^{\frac{1}{q_0}} \\ &\le C \int_{\mathbb{R}^n} Rf(x) R'h(x)\omega(x)\,dx.\end{aligned}$$

Again by Hölder's inequality and since R is bounded on $L_{p(\omega)}$ while R' is bounded on $L_q(\omega)$, we have

$$\begin{aligned}\|Tf\|_{L_p(\omega)} &\le C\left(\int_{\mathbb{R}^n} (Rf(x))^p \omega(x)\,dx\right)^{\frac{1}{p}} \left(\int_{\mathbb{R}^n} (R'h(x))^q \omega(x)\,dx\right)^{\frac{1}{q}} \\ &\le C\left(\int_{\mathbb{R}^n} |f(x)|^p \omega(x)\,dx\right)^{\frac{1}{p}} \left(\int_{\mathbb{R}^n} (h(x))^q \omega(x)\,dx\right)^{\frac{1}{q}} \\ &\le C\left(\int_{\mathbb{R}^n} |f(x)|^p \omega(x)\,dx\right)^{\frac{1}{p}},\end{aligned}$$

as needed. □

4.6 Fefferman's inequality

In his 1983 paper [36], Charles Fefferman proved the inequality

$$\int_{\mathbb{R}^n} |u(x)|^p |V(x)|\,dx \le C \int_{\mathbb{R}^n} |\nabla u(x)|^p\,dx \tag{4.29}$$

for all $u \in C_c^\infty$ and the potential V belonging to an appropriate space. He proved it for $p = 2$ assuming the potential $V \in L^{r,n-2r}$, with $1 < r \leq \frac{n}{2}$, where $L^{r,n-2r}(\mathbb{R}^n)$ is the classical Morrey space of the $L^r_{\mathrm{loc}}(\mathbb{R}^n)$ functions such that

$$\sup_{\substack{x\in\mathbb{R}^n,\\ \rho>0}} \frac{1}{\rho^{n-2r}} \int_{B(x,\rho)} |V(y)|^r \, dy < \infty,$$

where $B(x,\rho) = \{z \in \mathbb{R}^n \mid \|z - x\| < \rho\}$.

In a latter work, Chiarenza and Frasca [25] extended Fefferman's result with a different proof, assuming the potential V in $L^{r,n-pr}$ with $1 < r \leq \frac{n}{p}$ and $1 < p < n$.

Danielli, Garofallo, and Nhice [32] introduced a suitable version of Morrey spaces adapted to the Carnot–Carathéodory metric, and proved inequality (4.29) for V belonging to the Morrey space $L^{1,\lambda}$ for $\lambda > 0$.

Schecter [63] followed a different approach to inequality (4.29). He proved (4.29) for $p = 2$ and V in the Stummel–Kato class.

At the beginning of the twenty-first century, in [67], inequality (4.29) was proved with $1 < p < n$ and V in a more general class of potentials, namely nonlinear Kato class (for details in this class, see [9]). In [34], inequality (4.29) was proved by replacing the gradient on the right-hand side of (4.29) by an energy associated to an arbitrary system of vector fields, and the function V was taken from an appropriate Stummel–Kato class, defined via the Carnot–Carathéodory metric associated to the vector fields in a metric space.

In [17], inequality (4.29) was proved allowing $V \in A_1 \cap L_{\frac{n}{p}} \cap C_c^2$ with $1 < p < \frac{n}{p}$.

Taking into account these studies of Fefferman's inequality on diverse function spaces, it seems natural to study this inequality when the potential V belongs to a weighted Lebesgue space. To the best of our knowledge, this has not been done, yet. So, in this section we shall prove (4.29) allowing $V \in L_p(\omega)$ with $1 < p < \infty$.

This section is based on the paper "*The Fefferman inequality on weighted Lebesgue space $L_p(\omega)$*", by the first and second authors of this book, and it is currently under review for possible publication.

We begin by stating and proving the following lemma.

Lemma 4.2. *Let $B(x,r) = \{y \in \mathbb{R}^n : |y - x| < r\}$ denote the open ball of radius r and center at $x \in \mathbb{R}^n$. Then*

$$B(x,r) = \bigcup_{k=0}^{\infty} \{y : 2^{-k-1} r \leq |x - y| < 2^{-k} r\}.$$

Proof. Let $z \in B(x,r)$, then $|x - z| < r$. Since $\overline{\mathbb{Q}} = \mathbb{R}$, there exists $2^{-k} r \in \mathbb{Q}$ such that

$$|x - z| < 2^{-k} r < r.$$

Thus

$$|x - z| < 2^{-k}r. \tag{4.30}$$

Next, taking into account that $|x - z|$ and r are positive numbers, by the Archimedean property, there exists $2^{k+1} \in \mathbb{N}$ with $k \in \mathbb{Z}_+$ such that

$$r \leq 2^{k+1}|x - z|,$$

hence

$$2^{k+1}r \leq |x - z|. \tag{4.31}$$

Combining (4.30) and (4.31), we arrive at

$$2^{-k-1}r \leq |x - z| < 2^{-k}r,$$

which means that

$$z \in \bigcup_{k=1}^{\infty}\{y : 2^{-k-1}r \leq |x - z| < 2^{-k}r\}.$$

So far, we have proved that

$$B(x, r) \subset \bigcup_{k=1}^{\infty}\{y : 2^{-k-1}r \leq |x - z| < 2^{-k}r\}. \tag{4.32}$$

Next, let $z \in \bigcup_{k=0}^{\infty}\{y : 2^{-k-1}r \leq |x - z| < 2^{-k}r\}$. Then there exists $k_0 \in \mathbb{N}$ such that

$$2^{-k_0-1}r \leq |x - z| < 2^{-k_0}r,$$

from which we derive that

$$|x - z| < 2^{-k_0}r < r,$$

and thus

$$z \in B(x, r). \tag{4.33}$$

Finally, (4.32) and (4.33) tell us that

$$B(x, r) = \bigcup_{k=0}^{\infty}\{y : 2^{-k-1}r \leq |x - z| < 2^{-k}r\},$$

as needed. □

Corollary 4.1.

$$\{y : |x - y| \geq r\} = \mathbb{R}^n \setminus B(x,r) = \bigcup_{k=1}^{\infty} \{y : r2^{k-1} \leq |y - x| < r2^k\}.$$

The following result is the cornerstone in the proof of the main result of this section.

Lemma 4.3. *Let $u \in C_c^1(\mathbb{R}^n)$ and suppose that u and its partial derivatives of first order are integrable on $\mathbb{R}^n$. Then*

$$|u(x)| \leq \frac{1}{n\omega_n} \int_{\mathbb{R}^n} \frac{|\nabla u(y)|}{|x - y|^{n-1}} \, dy$$

for $x \in \mathbb{R}^n$, where ω_n is the Lebesgue measure of the unit ball in $\mathbb{R}^n$.

Proof. First, observe that

$$\frac{(x - y) \cdot \nabla u(y)}{|x - y|^n}$$

is integrable on $\mathbb{R}^n$ as a function of y. Actually, for $r > 0$, we have

$$\begin{aligned}\int_{\mathbb{R}^n} \frac{|(x - y) \cdot \nabla u(y)|}{|x - y|^n} \, dy &\leq \int_{B_r(x)} \frac{|\nabla u(y)|}{|x - y|^{n-1}} \, dy + \int_{\mathbb{R}^n \setminus B_r(x)} \frac{|\nabla u(y)|}{|x - y|^{n-1}} \, dy \\ &\leq \sup_{y \in B_r(x)} |\nabla u(y)| \int_{B_r(x)} \frac{dy}{|x - y|^{n-1}} + \frac{1}{r^{n-1}} \int_{\mathbb{R}^n} |\nabla u(y)| \, dy < \infty.\end{aligned}$$

Next, since $u \in C_c^1(\mathbb{R}^n)$, we also have

$$u(x) = -\int_0^\infty \frac{\partial}{\partial r} u(x + rz) \, dr, \tag{4.34}$$

where $z \in S^{n-1} = \{x \in \mathbb{R}^n : |x| = 1\}$. Integrating (4.34) over the whole unit sphere S^{n-1} yields

$$\begin{aligned}\omega_{n-1} u(x) &= \int_{S^{n-1}} u(x) \, d\sigma(z) \\ &= -\int_{S^{n-1}} \int_0^\infty \frac{\partial}{\partial r} u(x + rz) \, dr d\sigma(z) \\ &= -\int_{S^{n-1}} \int_0^\infty \nabla u(x + rz) \cdot z \, dr d\sigma(z) \\ &= -\int_0^\infty \int_{S^{n-1}} \nabla u(x + rz) \cdot z \, d\sigma(z) dr.\end{aligned}$$

Changing variables $y = x + rz$, $d\sigma(z) = r^{n-1} \, d\sigma(y)$, and

$$z = \frac{y-x}{|x-y|}, \qquad r = |x-y|,$$

we get

$$\begin{aligned}\omega_{n-1}u(x) &= -\int_0^\infty \int_{\partial B(x,r)} \nabla u(y)\cdot \frac{y-x}{|x-y|^n}\, d\sigma(y)dr\\ &= \int_{\mathbb{R}^n} \nabla u(y)\cdot \frac{x-y}{|x-y|^n}\, dy,\end{aligned}$$

which implies that

$$|u(x)| \le \frac{1}{n\omega_n}\int_{\mathbb{R}^n} \frac{|\nabla u(y)|}{|x-y|^{n-1}}\, dy,$$

as required. □

Theorem 4.10 (Fefferman's inequality). *Let $\Omega \subset \mathbb{R}^n$ be a bounded set. Consider $\omega \in A_p$ for any $u \in C_c^1(\mathbb{R}^n)$. If $V \in L_p(\omega)$ for $1 \le p < \infty$, then the following inequality holds:*

$$\int_\Omega |V(x)||u(x)|^p \omega(x)\, dx \le C(n,p,r)\left[\frac{[\omega]_p^p}{\omega(B)}\right]^{\frac{1}{p}} \|V\|_{L_{p(\omega)}} \int_\Omega |\nabla u(x)|^p \omega(x)\, dx$$

for any ball B in $\mathbb{R}^n$ and $C(n,p,r) = (2C_n C(n)m(B)r^{n-1})^p$.

Proof. For any $u \in C_c^1(\mathbb{R}^n)$, let us consider an open ball $B = B(x_0, r)$ such that $u \in C_c^1(B)$. By Lemma 4.3, we have for $C_n = \frac{1}{dn\omega_n}$,

$$\begin{aligned}|u(x)| &\le C_n \int_{\mathbb{R}^n} \frac{|\nabla u(y)|}{|x-y|}\, dy\\ &= C_n \int_{B(x,r)} \frac{|\nabla u(y)|}{|x-y|^{n-1}}\, dy + C_n \int_{\mathbb{R}^n\setminus B(x,r)} \frac{|\nabla u(y)|}{|x-y|^{n-1}}\, dy\\ &= A + B. \end{aligned}\tag{4.35}$$

In the sequel, we shall apply Lemmas 4.2, 4.3, Corollary 4.1, and Hölder's inequality.

For A, we have

$$\begin{aligned}A &= C_n \sum_{k=0}^\infty \int_{\{y:\frac{r}{2^{k+1}}\le|x-y|<\frac{r}{2^k}\}} \frac{|\nabla u(y)|}{|x-y|^{n-1}}\, dy\\ &\le C_n \sum_{k=0}^\infty \left(\frac{r}{2^{k+1}}\right)^{n-1} \int_{B(x,\frac{r}{2^k})} |\nabla u(y)|\, dy\\ &= C_n\left(\frac{r}{2}\right)^{n-1} \sum_{k=0}^\infty \left(\frac{1}{2^k}\right)^{n-1} \int_{B(x,\frac{r}{2^k})} |\nabla u(y)|\omega^{\frac{1}{p}}(y)\omega^{-\frac{1}{p}}(y)\, dy\end{aligned}$$

$$\le C_n\left(\frac{r}{2}\right)^{n-1}\sum_{k=0}^{\infty}\left(\frac{1}{2^k}\right)^{n-1}\left(\int_{B(x,\frac{r}{2^k})}|\nabla u(y)|^p\omega(y)\,dy\right)^{\frac{1}{p}}\left(\int_{B(x,\frac{r}{2^k})}\omega^{-\frac{q}{p}}(y)\,dy\right)^{\frac{1}{q}}$$

$$\le C_n\left(\frac{r}{2}\right)^{n-1}\sum_{k=0}^{\infty}\left(\frac{1}{2^k}\right)^{n-1}\left(\int_{\Omega}|\nabla u(y)|^p\omega(y)\,dy\right)^{\frac{1}{p}}\left(\int_{B(x,r)}\omega^{-\frac{q}{p}}(y)\,dy\right)^{\frac{1}{q}}. \tag{4.36}$$

Since $n \ge 1$, the series

$$\sum_{k=0}^{\infty}\left(\frac{1}{2^k}\right)^{n-1}$$

converges, say

$$\sum_{k=0}^{\infty}\left(\frac{1}{2^k}\right)^{n-1} = C(n) < \infty.$$

Hence, we can write (4.36) as

$$C_n\left(\frac{r}{2}\right)^{n-1}C(n)\left(\int_{\Omega}|\nabla u(y)|^p\omega(y)\,dy\right)^{\frac{1}{p}}\left(\int_{B(x,r)}\omega^{-\frac{q}{p}}(y)\,dy\right)^{\frac{1}{q}}.$$

Now, since $\frac{r}{2} \le r$,

$$\begin{aligned}A &= C_nC(n)\left(\frac{r}{2}\right)^{n-1}\left(\int_{\Omega}|\nabla u(y)|^p\omega(y)\,dy\right)^{\frac{1}{p}}\left(\int_{B(x,r)}\omega^{-\frac{q}{p}}(y)\,dy\right)^{\frac{1}{q}}\\ &\le C_nC(n)r^{n-1}\left(\int_{\Omega}|\nabla u(y)|^p\omega(y)\,dy\right)^{\frac{1}{p}}\left(\int_{B(x,r)}\omega^{-\frac{q}{p}}(y)\,dy\right)^{\frac{1}{q}}.\end{aligned}$$

Thus

$$A \le C_nC(n)r^{n-1}\left(\int_{\Omega}|\nabla u(y)|^p\omega(y)\,dy\right)^{\frac{1}{p}}\left(\int_{B(x,r)}\omega^{-\frac{q}{p}}(y)\,dy\right)^{\frac{1}{q}}. \tag{4.37}$$

For B, we obtain

$$\begin{aligned}B &= C_n\sum_{k=0}^{\infty}\int_{\{y:r2^k\le|x-y|<r2^{k+1}\}}\frac{|\nabla u(y)|}{|x-y|^{n-1}}\,dy\\ &\le C_n\sum_{k=0}^{\infty}\left(\frac{1}{r2^k}\right)^{n-1}\int_{B(x,r2^k)}|\nabla u(y)|\,dy\\ &\le \frac{C_n}{r^{n-1}}\sum_{k=0}^{\infty}\left(\frac{1}{2^k}\right)^{n-1}\left(\int_{\Omega}|\nabla u(y)|^p\omega(y)\,dy\right)^{\frac{1}{p}}\left(\int_{B(x,r)}\omega^{-\frac{q}{p}}(y)\,dy\right)^{\frac{1}{q}}.\end{aligned}$$

Hence

$$B \le C_n C(n) r^{1-n} \Big(\int_\Omega |\nabla u(y)|^p \omega(y)\, dy \Big)^{\frac{1}{p}} \Big(\int_{B(x,r)} \omega^{-\frac{q}{p}}(y)\, dy \Big)^{\frac{1}{q}}. \tag{4.38}$$

Inequalities (4.37), (4.38), and (4.35) together imply

$$|u(x)| \le 2C_n C(n) r^{1-n} \Big(\int_\Omega |\nabla u(y)|^p \omega(y)\, dy \Big)^{\frac{1}{p}} \Big(\int_{B(x,r)} \omega^{-\frac{q}{p}}(y)\, dy \Big)^{\frac{1}{q}},$$

and so

$$|u(x)|^p \le (2C_n C(n) r^{1-n})^p \Big(\int_\Omega |\nabla u(y)|^p \omega(y)\, dy \Big) \Big(\int_{B(x,r)} \omega^{-\frac{q}{p}}(y)\, dy \Big)^{p-1}. \tag{4.39}$$

Next, multiplying both sides of (4.39) by $|V(x)|$, integrating with respect to $\omega(x)\,dx$, and invoking one more time Hölder's inequality, we obtain

$$\begin{aligned}
&\int_\Omega |V(x)||u(x)|^p \omega(x)\, dx \\
&= \int_{B(x_0,r)} |V(x)||u(x)|^p \omega(x)\, dx \\
&\le (2C_n C(n) r^{1-n})^p \int_{B(x_0,r)} |V(x)|\omega(x)\, dx \Big(\int_\Omega |\nabla u(y)|^p \omega(y)\, dy \Big) \Big(\int_{B(x,r)} \omega^{-\frac{q}{p}}(y)\, dy \Big)^{p-1} \\
&= (2C_n C(n) r^{1-n})^p \Big(\int_\Omega |V(x)|^p \omega(x)\, dx \Big)^{\frac{1}{p}} \Big(\int_{B(x_0,r)} \omega(x)\, dx \Big)^{p-1} \\
&\quad \times \Big(\int_{B(x_0,r)} \omega^{-\frac{q}{p}}(x)\, dx \Big)^{p-1} \Big(\int_\Omega |\nabla u(y)|^p \omega(y)\, dy \Big) \\
&= \frac{(2C_n C(n) r^{1-n})^p \|V\|_{L_p(\omega)}}{(\omega(B))^{\frac{1}{p}}} \\
&= \frac{(2C_n C(n) r^{1-n})^p (\mathrm{m}(B))^p}{(\omega(B))^{\frac{1}{p}}} \Big(\frac{1}{\mathrm{m}(B)} \int_B \omega(x)\, dx \Big) \\
&\quad \times \Big(\frac{1}{\mathrm{m}(B)} \int_B \omega^{-\frac{q}{p}}(x)\, dx \Big)^{p-1} \|V\|_{L_p(\omega)} \Big(\int_\Omega |\nabla u(y)|^p \omega(y)\, dy \Big) \\
&\le \frac{(2C_n C(n) r^{1-n} \mathrm{m}(B))^p}{(\omega(B))^{\frac{1}{p}}} [\omega]_p \|V\|_{L_p(\omega)} \int_\Omega |\nabla u(y)|^p \omega(y)\, dy.
\end{aligned}$$

Finally, we arrived at

$$\int_\Omega |V(x)||u(x)|^p \omega(x)\, dx \le C(n,r,p) \Big[\frac{[\omega]_p^p}{\omega(B)} \Big]^{\frac{1}{p}} \|V\|_{L_p(\omega)} \int_\Omega |\nabla u(x)|^p \omega(x)\, dx,$$

where $C(n,r,p) = (2C_n C(n) \mathrm{m}(B) r^{1-n})^p$. □

5 Extensions of weighted Lebesgue spaces: Lambda and Gamma spaces

In mathematics, it is always interesting to generalize concepts and then try to extend the known results to the new setting. In this regard, we present two sets which may be seen as generalizations of weighted $L_p(\omega)$ spaces, and we also study some of their more relevant properties.

Let us begin by recalling two important concepts: *distributions and decreasing rearrangements of functions.*

5.1 Distributions and decreasing rearrangements of functions

Let f be a complex-valued measurable function defined on a σ-finite measure space $(X, \mathcal{A}, \mu)$. For $\lambda \geq 0$, the *distribution function* of f is defined as

$$D_f(\lambda) = \mu\{x \in X : |f(x)| > \lambda\}. \tag{5.1}$$

Observe that D_f depends only on the absolute value $|f|$ of the function f and D_f may take the value $+\infty$.

The distribution function D_f provides information about the size of f, but not about the behavior of f itself near any given point. For instance, a function on $\mathbb{R}^n$ and each of its translates have the same distribution function. It follows from (5.1) that D_f is a decreasing function of λ (not necessarily strictly) and continuous from the right.

By f^* we mean the *decreasing rearrangement* of f given as

$$f^*(t) = \inf\{\lambda > 0 : D_f(\lambda) \leq t\}, \quad t > 0, \tag{5.2}$$

where we use the convention that $\inf \emptyset = +\infty$. This rearrangement f^* is decreasing and right-continuous. Notice that

$$f^*(0) = \inf\{\lambda > 0 : D_f(\lambda) \leq 0\} = \|f\|_\infty,$$

since

$$\|f\|_\infty = \inf\{\alpha \geq 0 : \mu(\{x \in X : |f(x)| > \alpha\}) = 0\}.$$

Also observe that if D_f is strictly decreasing, then

$$f^*(D_f(t)) = \inf\{\lambda > 0 : D_f(\lambda) \leq D_f(t)\} = t.$$

This proves that f^* is the inverse function of the distribution function D_f.

https://doi.org/10.1515/9783112223246-005

Let $\mathcal{F}(X,\mathcal{A})$ denote the set of all $\mathcal{A}$-measurable functions on X. Let $(X,\mathcal{A},\mu)$ and $(Y,\mathcal{B},\nu)$ be two measure spaces.

Two functions $f\in\mathcal{F}(X,\mathcal{A})$ and $g\in\mathcal{F}(Y,\mathcal{B})$ are said to be equimeasurable if they have the same distribution function, that is, if

$$\mu(\{x\in X:|f(x)|>\lambda\})=\nu(\{y\in Y:|g(y)|>\lambda\}),\quad \text{for all } \lambda\geq 0. \tag{5.3}$$

So then there exists a right-continuous decreasing function f^* equimeasurable with f. Hence the decreasing rearrangement is unique.

In what follows, we gather some useful properties of the distribution function and the decreasing rearrangement function: For $\lambda_1,\lambda_2\geq 0$,

1. $|g|\leq|f|$ μ-a. e. implies that $D_g\leq D_f$.
2. $D_{cf}(\lambda)=D_f(\frac{\lambda}{|c|})$ for all $c\in\mathbb{C}\setminus\{0\}$.
3. $D_{f+g}(\lambda_1+\lambda_2)\leq D_f(\lambda_1)+D_g(\lambda_2)$.
4. $D_{fg}(\lambda_1\lambda_2)\leq D_f(\lambda_1)+D_g(\lambda_2)$.
5. $f^*(t)>\lambda$ if and only if $D_f(\lambda)>t$.
6. If $|f|\leq\liminf_{n\to\infty}|f_n|$, then $f^*\leq\liminf_{n\to\infty}f_n^*$.
7. If $E\in\mathcal{A}$, then $(\mathcal{X}_E)^*(t)=\mathcal{X}_{[0,\mu(E))}(t)$.
8. If $E\in\mathcal{A}$, then $(f\,\mathcal{X}_E)^*(t)\leq f^*(x)\,\mathcal{X}_{[0,\mu(E))}(t)$.

For more details on distribution functions and decreasing rearrangements, see [2, 16, 40].

The following inequality is sometimes called *Hardy–Littlewood inequality*. A proof of it may be found in [2, Theorem 2.2].

Lemma 5.1. *Let $(X,\mathcal{A},\mu)$ be a σ-finite measure space. Let $f,g\in\mathcal{F}(X,\mathcal{A})$, then*

$$\int_X |fg|\,d\mu\leq\int_0^\infty f^*(t)g^*(t)\,dt.$$

Lemma 5.2. *Let $(X,\mathcal{A},\mu)$ be a σ-finite nonatomic measure space. Suppose f belongs to $\mathcal{F}(X,\mathcal{A})$ and $0<t<\mu(X)$. Then there exists a measurable set E_t with $\mu(E_t)=t$, such that*

$$\int_{E_t}|f|\,d\mu=\int_0^t f^*(s)\,ds.$$

For a proof of the above result, see [2, Lemma 2.5].

As an immediate consequence of Lemma 5.2, we obtain that, for a σ-finite nonatomic measure space $(X,\mathcal{A},\mu)$ and $t\in(0,\mu(X))$, we can construct a set A with $\mu(A)=t$.

Definition 5.1. A σ-finite measure space $(X,\mathcal{A},\mu)$ is said to be resonant if for every pair $f,g\in\mathcal{F}(X,\mathcal{A})$, we have

$$\sup_{\overline{g}=g} \int_X |f\overline{g}| d\mu = \int_0^\infty f^*(t) g^*(t) dt,$$

for all functions $\overline{g}$ equimeasurable with g, that is, $D_{\overline{g}}(\lambda) = D_g(\lambda)$, $\lambda \geq 0$.

Note that $\mathfrak{L}$ denotes the Lebesgue σ-algebra and m stands for the Lebesgue measure.

Remark 5.1. The measure space $((0, +\infty), \mathfrak{L}, m)$ is resonant.

With the aim of illustrating how the calculations are done in this latter space, let us consider the function $f : (0, +\infty) \to \mathbb{R}$ given by $f(x) = 1 - e^{-x}$. Observe that

$$\begin{aligned} D_f(\lambda) &= m(\{x \in (0, +\infty) : |f(x)| > \lambda) \\ &= \begin{cases} +\infty & \text{if } 0 \leq \lambda < 1, \\ 0 & \text{if } \lambda \geq 1. \end{cases} \end{aligned}$$

Since $f^*(t) = \inf\{\lambda > 0 : D_f(\lambda) \geq t\}$, we have $f^*(t) = 1$. Let $g = \mathcal{X}_{[0,1]}$ and $\overline{g} = \mathcal{X}_{[1,2]}$. Note that $D_g(\lambda) = D_{\overline{g}}(\lambda)$ and $g^*(t) = \overline{g}^*(t) = \mathcal{X}_{[0,1]}(t)$. On the one hand,

$$\begin{aligned} \int_0^{+\infty} |f(x)\overline{g}(x)|\, dx &= \int_0^{+\infty} (1 - e^{-x}) \mathcal{X}_{[1,2]}(x)\, dx \\ &= \int_1^2 (1 - e^{-x})\, dx \\ &= 1 - \frac{e-1}{e} < 1. \end{aligned}$$

On the other hand,

$$\int_0^{+\infty} f^*(t) g^*(t)\, dt = \int_0^{+\infty} \mathcal{X}_{[0,1]}(t)\, dt = 1.$$

Therefore,

$$\sup_{\overline{g} \sim g} \int_0^{+\infty} |f(t)\overline{g}(t)|\, dt = 1 = \int_0^{+\infty} f^*(t) g^*(t)\, dt.$$

Remark 5.2. Not every measure space is resonant.

A typical example of a nonresonant measure space is an atomic measure space having at least two atoms of unequal measure. More precisely, let $a, b \in X$ be such that $\mu(\{a\}) = \alpha$, $\mu(\{b\}) = \beta$ with $0 < \alpha < \beta$. Let $f = \mathcal{X}_{\{b\}}$ and $g = \mathcal{X}_{\{a\}}$. Then it is clear that every function $\overline{g}$ satisfying $g \sim \overline{g}$ must obey $b \notin \operatorname{supp} \overline{g}$. Therefore, for every such $\overline{g}$, we have $\int_X |f\overline{g}|\, d\mu = 0$. On the other hand, $f^* = \mathcal{X}_{[0,\beta]}$ and $g^* = \mathcal{X}_{[0,\alpha]}$, hence $\int_0^{+\infty} f^*(t) g^*(t)\, dt = \alpha$. Consequently, no such measure space can be resonant.

Lemma 5.3. *Every nonatomic σ-finite measure space is resonant.*

For a proof of the above lemma, see [2, Theorem 2.7].

As in the previous chapters, q stands for the conjugate exponent of p, i. e.,

$$\frac{1}{p} + \frac{1}{q} = 1, \quad \text{for } p \in (1, \infty).$$

Let A be a set and $F, G : A \to [0, \infty)$ be two functions. We say F and G are equivalent, using the notation

$$F \approx G,$$

if there exists a constant $C \in [1, \infty)$ such that

$$C^{-1} F(x) \le G(x) \le C F(x)$$

for all $x \in A$.

Now, we need to recall some propositions and definitions from the theory of Banach function spaces, which will be useful in the sequel.

Definition 5.2. Let $\mathcal{B}$ be Banach function space. The associate space is the space of all measurable functions g satisfying

$$\|g\|_{X'} = \sup_{\|f\|_X \le 1} \int_X |fg| d\mu < \infty.$$

The associate norm is quite difficult to investigate from the expression in Definition 5.2. So we will find a formula to express it with an equivalent norm.

Remark 5.3. Let $(X, \mathcal{A}, \mu)$ be a nonatomic σ-finite measure space.

If we consider a rearrangement invariant Banach function space (that is, a Banach function space $\mathcal{B}$, for which $\|f\|_{\mathcal{B}} = \|g\|_{\mathcal{B}'}$ for every pair of equimeasurable functions, $D_f(\lambda) = D_g(\lambda)$, $\forall \lambda \ge 0$), then we may rewrite its associate norm as follows:

$$\|g\|_{\mathcal{B}'} = \frac{\int_X |fg| d\mu}{\|f\|_{\mathcal{B}}} = \frac{\int_0^\infty f^*(t) g^*(t) dt}{\|f\|_{\mathcal{B}}}.$$

We point out that one of the most important properties of f^* is that

$$\|f\|_p = \left(\int_X |f|^p d\mu \right)^{\frac{1}{p}} = \left(\int_0^\infty (f^*(t))^p dt \right)^{\frac{1}{p}},$$

which is obtained from the fact that f and f^* are equimeasurable. This allows us to study L_p spaces via decreasing reordering.

5.2 Lambda spaces

Recall that by a weight function w we mean a function $w : (0, \infty) \to (0, \infty)$ which is locally integrable and such that

$$\int_0^\infty w(t) dt < \infty.$$

The Lorentz spaces are a two-parameter family of function spaces, $L(p, q)$, which generalize the Lebesgue space L_p. The *Lorentz spaces* $L(p, q)$ were introduced by G. G. Lorentz in [54, 55]. A general treatment of Lorentz spaces is given in the article of Hunt [47]. Let $\mathfrak{F}(\mathfrak{X}, \mathcal{A})$ denote the set of all $\mathcal{A}$-measurable functions on $\mathfrak{X}$. Next, for any $f \in \mathfrak{F}(\mathfrak{X}, \mathcal{A})$, and any two extended real numbers p and q in the set $[1, \infty]$, we write

$$\|f\|_{(p,q)} = \left(\int_0^\infty (t^{1/p} f^*(t))^q \frac{dt}{t} \right)^{1/q} \quad \text{if } q < \infty. \tag{5.4}$$

Thus $\|\cdot\|_{(p,q)}$ are extended nonnegative functions on $\mathfrak{F}(\mathfrak{X}, \mathcal{A})$, and the Lorentz spaces will be defined in terms of the $\|\cdot\|_{(p,q)}$ norms. If $(\mathfrak{X}, \mathcal{A}, \mu)$ is a σ-finite measure space, w is a weight in $(0, \infty)$, and $0 < p < \infty$, the Lorentz space $\Lambda^p_{\mathfrak{X}}(w)$ (or just *Lambda space*) is defined to be the class of all measurable functions f in $\mathfrak{X}$ for which

$$\|f\|_{\Lambda^p_{\mathfrak{X}}(w)} = \left(\int_0^\infty (f^*(t))^p w(t) dt \right)^{1/p} < \infty, \tag{5.5}$$

where f^* is the decreasing rearrangement of f with respect to μ. Theses spaces were first introduced by G. Lorentz in [54] for the case $\mathfrak{X} = (0, 1)$.

By choosing w properly, one obtains the space $L(p, q)$ defined in (5.4) as we shall see. In the next section, we will study analytical properties of these function spaces.

In this section, $(\mathfrak{X}, \mathcal{A}, \mu)$ denotes, except when otherwise mentioned, a general measure space. For a given weight w, we write

$$W(r) = \int_0^r w(t)dt < \infty,$$

with $0 \leq r < \infty$. If $d\mu(t) = w(t)dt$, we will write $D_f^W(\lambda)$ and $f_w^*(t)$ to show the dependence on the weight w. If $(\mathfrak{X}, \mathcal{A}, \mu) = (\mathbb{R}^n, \mathcal{B}, w(t)dt)$, we write $L_{(p,q)}(w)$ instead of $L_{(p,q)}(\mathfrak{X})$.

For more on Lorentz spaces, check references [1, 2, 7, 8, 12–15, 22, 40].

Definition 5.3. Let w be a weight on $(0, \infty)$. For $0 < p \leq \infty$, we define the functional $\|\cdot\|_{\Lambda^p_{\mathfrak{X}}(w)} : \mathfrak{F}(\mathfrak{X}, \mathcal{A}) \to [0, \infty]$ as

$$\|f\|_{\Lambda^p_{\mathfrak{X}}(w)} = \left(\int_0^\infty (f^*(t))^p w(t)dt \right)^{1/p}.$$

The Lorentz space $\Lambda^p(w) = \Lambda^p_{\mathfrak{X}}(w)$ is the class

$$\Lambda^p_{\mathfrak{X}}(w) = \{ f \in \mathfrak{F}(\mathfrak{X}, \mathcal{A}) : \|f\|_{\Lambda^p_{\mathfrak{X}}(w)} < \infty \}.$$

Observe that $\|f\|_{\Lambda^p_{\mathfrak{X}}(w)} = \|f^*\|_{L_p(w)}$. From this, we can extend the previous definition for

$$\Lambda^{p,q}_{\mathfrak{X}}(w) = \{ f \in \mathfrak{F}(\mathfrak{X}, \mathcal{A}) : \|f\|_{\Lambda^{(p,q)}_{\mathfrak{X}}} = \|f^*\|_{L_{(p,q)}(w)} < \infty \}.$$

Remark 5.4. If $0 < p, q < \infty$, $\Lambda^q_{\mathfrak{X}}(t^{\frac{q}{p}-1}) = L_{(p,q)}(t^{\frac{q}{p}-1})$ and in this case $W(t) = \frac{p}{q} t^{\frac{q}{p}}$, $t \geq 0$.

Proposition 5.1. *Let $(\mathfrak{X}, \mathcal{A}, \mu)$ be a σ-finite measure space. For $0 < p < \infty$,*

$$\|f\|_{\Lambda^p_{\mathfrak{X}}(w)} = \left(\int_0^\infty p t^{p-1} W(D_f(t)) dt \right)^{1/p}.$$

Proof. We can write

$$\begin{aligned} \|f\|_{\Lambda^p_{\mathfrak{X}}(w)} &= \left(\int_0^\infty (f^*(t))^p w(t)\, dt \right)^{1/p} \\ &= \left(\int_0^\infty \left(\int_0^{f^*(t)} p\lambda^{p-1}\, d\lambda \right) w(t)\, dt \right)^{1/p} \\ &= \left(\int_0^\infty \left(\int_0^\infty p\lambda^{p-1} \chi_{(0,f^*(t))}(\lambda)\, d\lambda \right) w(t)\, dt \right)^{1/p} \end{aligned}$$

$$= \left(\int_0^\infty \left(\int_0^\infty p\lambda^{p-1} \chi_{\{\lambda>0: f^*(t)>\lambda\}}(\lambda)\, d\lambda \right) w(t)\, dt \right)^{1/p}$$

$$= \left(\int_0^\infty \left(\int_0^\infty p\lambda^{p-1} \chi_{\{t>0: D_f(\lambda)>t\}}(t)\, d\lambda \right) w(t)\, dt \right)^{1/p}.$$

Applying Fubini's theorem, we have

$$\left(\int_0^\infty \left(\int_0^\infty p\lambda^{p-1} \chi_{\{t>0: D_f(\lambda)>t\}}(t)\, d\lambda \right) w(t)\, dt \right)^{1/p}$$

$$= \left(\int_0^\infty p\lambda^{p-1} \left(\int_0^\infty \chi_{\{t>0: D_f(\lambda)>t\}}(t) w(t)\, dt \right) d\lambda \right)^{1/p}$$

$$= \left(\int_0^\infty p\lambda^{p-1} \left(\int_0^{D_f(\lambda)} w(t)\, dt \right) d\lambda \right)^{1/p}$$

$$= \left(\int_0^\infty p\lambda^{p-1} W(D_f(\lambda))\, d\lambda \right)^{1/p}.$$

□

The following result gives several equivalent expressions for the functional $\|\cdot\|_{\Lambda^{p,q}_{\mathfrak{X}}(w)}$. In particular, we see that it only depends on W.

Proposition 5.2. *Let $(\mathfrak{X}, \mathcal{A}, \mu)$ be a measure space. For $0 < p, q < \infty$ and $f \in \mathfrak{F}(\mathfrak{X}, \mathcal{A})$:*
(i) $\|f\|_{\Lambda^{p,q}_{\mathfrak{X}}(w)} = (\int_0^\infty pt^{q-1}(W(D_f(t)))^{q/p} dt)^{1/q}$,
(ii) $\|f\|_{\Lambda^{p,\infty}_{\mathfrak{X}}(w)} = \sup_{t>0} t(W(D_f(t)))^{1/p} = \sup_{t>0} f^*(t)(W(t))^{1/p}$.

Proof. (i) Since $D_f(t) = D_{f^*}(t)$ (see [2]), we have

$$W(D_f(t)) = W(D_{f^*}(t)) = \int_0^{D_{f^*}(t)} w(s)ds$$

$$= \int_0^\infty \mathcal{X}_{(0, D_{f^*}(t))}(s) w(s) ds = \int_0^\infty (\mathcal{X})^*_{\{f^*(s)>t\}}(s) w(s) ds$$

$$= \int_{\{f^*(s)>t\}} w(s)ds$$

$$= D^W_{f^*}(t).$$

Also, note that $D^W_{f^*}(t) = (f^*)^*_w(t)$. Thus we obtain

$$\left(\int_0^\infty pt^{q-1}[W(D_f(t))]^{\frac{q}{p}}dt\right)^{1/q} = \left(\int_0^\infty pt^{q-1}(D^w_{f^*}(t))^{\frac{q}{p}}dt\right)^{1/q}$$
$$= \left(\int_0^\infty t^{\frac{q}{p-1}}\int_0^{(f^*)^*_w(t)} qs^{q-1}dsdt\right)^{1/q}$$
$$= \left(\int_0^\infty (t^{1/p}(f^*)^*_w(t))^q\frac{dt}{t}\right)^{1/q}$$
$$= \|f^*\|_{L_{(p,q)}(w)}$$
$$= \|f\|_{\Lambda^{p,q}_{\mathfrak{X}}(w)}.$$

(ii) One can observe that

$$\|f\|_{\Lambda^{p,q}_{\mathfrak{X}}(w)} = \|f^*\|_{L_{(p,\infty)}(w)} = \sup_{t>0} t(D^w_{f^*}(t))^{1/p} = \sup_{t>0} t(W(D_f(t)))^{1/p}$$
$$= \sup_{t>0} f^*(t)(W(t))^{1/p}.$$

□

Remark 5.5.

(a) Comparing Propositions 5.1 and 5.2(i), we see that for $q < \infty$, $\|f\|_{\Lambda^{p,q}_{\mathfrak{X}}(w)} = \|f\|_{\Lambda^q_{\mathfrak{X}}(w_0)}$, where $w_0(t) = (W(t))^{\frac{q}{p-1}}w(t)$ with $0 < t < \mu(\mathfrak{X})$. Therefore every Lorentz space as defined here reduces to $\Lambda^p_{\mathfrak{X}}(w)$ and its weak version $\Lambda^{p,\infty}_{\mathfrak{X}}(w)$.

(b) From Proposition 5.2(ii), we deduce that $\Lambda^{p,\infty}_{\mathfrak{X}}(w) = \Lambda^{q,\infty}_{\mathfrak{X}}(\frac{q}{p}w_0)$ for $0 < p, q < \infty$.

(c) Observe that (a) make sense because $f^*(t) = 0$ if $t \geq \mu(\mathfrak{X})$. Hence, the behavior of the weight w on $[\mu(\mathfrak{X}), \infty)$ is irrelevant.

For $L_{(p,\infty)}(\mathfrak{X})$, it is known that the quasinorm $\|f\|_{(p,\infty)}$ is, for every $q < p$, equivalent to the functional

$$\sup_{E\subset\mathfrak{X}} \|f\,\mathcal{X}_E\|_q(\mu(E))^{\frac{1}{p}-\frac{1}{q}}.$$

This is the so-called Kolmogorov condition (see [37]). An analogous version for $\Lambda^{p,\infty}_{\mathfrak{X}}(w)$ also holds.

Proposition 5.3 (Chebyschev-type inequality). *Let $f \in \Lambda^p_{\mathfrak{X}}(w)$, then*

$$W(D_f(t)) \leq \frac{\|f\|^p_{\Lambda^p_{\mathfrak{X}}(w)}}{t^p}.$$

Proof. Let $E = \{x \in \mathfrak{X} : |f(x)| > t|\}$, then observe that $t\mathcal{X}_E < |f(x)|$. For this, we have

$$t^p\mathcal{X}_{(0,\mu(E))}(t) < (f^*(t))^p.$$

Thus

$$t^p \int_0^{\mu(E)} w(t)dt \le \int_0^{\infty} (f^*(t))^p w(t)dt.$$

Finally,

$$W(D_f(t)) \le \frac{\|f\|^p_{\Lambda^p_{\mathfrak{X}}(w)}}{t^p}. \qquad \square$$

Proposition 5.4. *If $0 < q < p < \infty$ and $f \in \mathfrak{F}(\mathfrak{X}, \mathcal{A})$, then*

$$\|f\|_{\Lambda^{p,\infty}_{\mathfrak{X}}(w)} \le \sup_{E \subset \mathfrak{X}} \|f \mathcal{X}_E\|_{\Lambda^q_{\mathfrak{X}}(w)} [W(\mu(E))]^{\frac{1}{p}-\frac{1}{q}} \le \left(\frac{p}{p-q}\right)^{1/q} \|f\|_{\Lambda^{p,\infty}_{\mathfrak{X}}(w)},$$

where the supremum is taken over all measurable sets $E \subset \mathfrak{X}$.

Proof. In order to prove the first inequality, let us consider the set E of $\mathfrak{X}$ given by

$$E = \{x \in \mathfrak{X} : |f(x)| > t\}$$

and also let

$$S = \sup_{E \subset \mathfrak{X}} \|f \mathcal{X}_E\|_{\Lambda^q_{\mathfrak{X}}(w)} (W(\mu(E)))^{\frac{1}{p}-\frac{1}{q}}.$$

Then

$$\begin{aligned}
S &\ge \|f \mathcal{X}_E\|_{\Lambda^q_{\mathfrak{X}}(w)} (W(\mu(E)))^{\frac{1}{p}-\frac{1}{q}} \\
&= \left(\int_0^{\infty} [(f \mathcal{X}_E)^*(s)]^q w(s)ds \right)^{1/q} (W(\mu(E)))^{\frac{1}{p}-\frac{1}{q}} \\
&\ge \left(t^q \int_0^{\mu(E)} w(s)ds \right)^{1/q} (W(\mu(E)))^{\frac{1}{p}-\frac{1}{q}} \\
&= t(W(\mu(E)))^{\frac{1}{q}} (W(\mu(E)))^{\frac{1}{p}-\frac{1}{q}} = t(W(\mu(E)))^{\frac{1}{p}} \\
&= t(W(D_f(t)))^{\frac{1}{p}}.
\end{aligned}$$

Taking the supremum over $t > 0$, we get

$$\|f\|_{\Lambda^{p,\infty}_{\mathfrak{X}}(w)} \le S. \tag{5.6}$$

To prove the second inequality, for each $f \in \mathfrak{F}(\mathfrak{X}, \mathcal{A})$, $E \subset \mathfrak{X}$, let $a = \|f\|_{\Lambda^{p,\infty}_{\mathfrak{X}}(w)}(W(\mu(E)))^{-\frac{1}{p}}$. Then

$$\|f\mathcal{X}_E\|_{\Lambda^q_{\mathfrak{X}}(w)} = \int_0^\infty qt^{q-1}W(D_{f\mathcal{X}_E}(t))dt$$
$$= \int_0^a qt^{q-1}W(D_{f\mathcal{X}_E}(t))dt + \int_a^\infty qt^{q-1}W(D_{f\mathcal{X}_E}(t))dt.$$

Note that due to

$$D_{f\mathcal{X}_E}(t) = \mu(\{x \in \mathfrak{X} : |f\mathcal{X}_E(x)| > t\}) \leq \mu(E)$$

and $f\mathcal{X}_E \leq f$, one has $D_{f\mathcal{X}_E}(t) \leq D_f(t)$, thus

$$\begin{aligned}\|f\mathcal{X}_E\|^q_{\Lambda^q_{\mathfrak{X}}(w)} &\leq W(\mu(E))\int_0^a qt^{q-1}dt + \int_a^\infty qt^{q-1}W(D_f(t))dt \\ &\leq W(\mu(E))a^q + \int_a^\infty qt^{q-1}\|f\|^p_{\Lambda^{p,\infty}_{\mathfrak{X}}(w)}\frac{dt}{t^p} \\ &= W(\mu(E))a^q + \frac{q}{p-q}\|f\|^p_{\Lambda^{p,\infty}_{\mathfrak{X}}(w)}a^{q-p} \\ &\leq \frac{q}{p-q}\|f\|^q_{\Lambda^{p,\infty}_{\mathfrak{X}}(w)}(W(\mu(E)))^{\frac{p-q}{p}}.\end{aligned}$$

Hence

$$\|f\mathcal{X}_E\|_{\Lambda^q_{\mathfrak{X}}(w)}(W(\mu(E)))^{\frac{1}{p}-\frac{1}{q}} \leq \left(\frac{p}{p-q}\right)^{\frac{1}{q}}\|f\|_{\Lambda^{p,\infty}_{\mathfrak{X}}(w)}. \qquad \square$$

In the following proposition, we state some elementary properties for these spaces.

Proposition 5.5. *For $0 < p < \infty$ and f, g, f_k, $k \geq 1$, functions belonging to $\mathfrak{F}(\mathfrak{X}, \mathcal{A})$, we have that*
(i) $|f| \leq |g|$ *implies* $\|f\|_{\Lambda^p_{\mathfrak{X}}(w)} \leq \|g\|_{\Lambda^p_{\mathfrak{X}}(w)}$;
(ii) $\|af\|_{\Lambda^p_{\mathfrak{X}}(w)} = |a|\|f\|_{\Lambda^p_{\mathfrak{X}}(w)}$;
(iii) *if* $0 \leq f_k \leq f_{k+1} \to f$ *a. e., then* $\lim_{k\to\infty}\|f_k\|_{\Lambda^p_{\mathfrak{X}}(w)} = \|f\|_{\Lambda^p_{\mathfrak{X}}(w)}$;
(iv) $\|\liminf |f_k|\|_{\Lambda^p_{\mathfrak{X}}(w)} \leq \liminf \|f_k\|_{\Lambda^p_{\mathfrak{X}}(w)}$;
(v) $\Lambda^q_{\mathfrak{X}}(w) \subset \Lambda^p_{\mathfrak{X}}(w)$ *for* $0 < p < q < \infty$, $W(\mu(\mathfrak{X})) < \infty$;
(vi) $\chi_E \in \Lambda^p_{\mathfrak{X}}(w)$ *if* $\mu(E) < \infty$.

Proof. Properties (i) and (ii) follow from the monotonicity of the rearrangement and Definition 5.3. To prove (iii), just observe that $0 \leq f_k \leq f_{k+1} \to f$ implies that

$$\lim_{k\to\infty} D_{f_k}(\lambda) = D_f(\lambda).$$

Next, let $F_k(t) = D_{f_k}(t)$. Then $f_k^*(t) = m(\{\lambda > 0 : D_{f_k}(\lambda) > t\}) = D_{F_k}(t)$. Since $D_{f_k}(t) \le D_{f_{k+1}}(t)$, we have $F_k(t) \le F_{k+1}(t)$. Thus $E_{F_1}(t) \subseteq E_{F_2}(t) \subseteq \cdots$, and then

$$E_F(t) = \bigcup_{k=1}^{\infty} E_{F_k}(t), \quad \text{where } E_{F_k}(t) = \{F_k > t\}.$$

Therefore $\lim_{k\to\infty} D_{F_k}(t) = D_F(t)$. Hence $\lim_{k\to\infty} f_k^*(t) = f^*(t)$.

By the monotone convergence theorem, we have

$$\lim_{k\to\infty} \|f_k\|_{\Lambda^p_{\mathfrak{X}}(w)} = \|f\|_{\Lambda^p_{\mathfrak{X}}(w)}.$$

(iv) The distribution function of $\liminf |f_n|$ satisfies

$$\begin{aligned}
\underset{\liminf |f_n|}{\mathrm{D}}(\lambda) &= \mu(\{x : \liminf |f_n(x)| > \lambda\}) \\
&= \mu(\liminf \{x : |f_n(x)| > \lambda\}) \\
&= \mu\Big(\bigcup_{n=1}\bigcap_{k=n}\{x : |f_k(x)| > \lambda\}\Big) \\
&= \liminf \mu(\{x : |f_n(x)| > \lambda\}) \\
&= \liminf D_{f_n}(\lambda) \le \liminf D_{|f_n|}(\lambda),
\end{aligned}$$

thus,

$$\underset{\liminf |f_n|}{\mathrm{D}}(\lambda) \le \liminf D_{|f_n|}(\lambda).$$

From this, we obtain

$$\begin{aligned}
\inf\Big\{\lambda > 0 : \underset{\liminf |f_n|}{\mathrm{D}}(\lambda) \le t\Big\} &\le \inf\Big\{\lambda > 0 : \liminf \underset{|f_n|}{\mathrm{D}}(\lambda) \le t\Big\} \\
&\le \liminf\Big(\inf\Big\{\lambda > 0 : \underset{|f_n|}{\mathrm{D}}(\lambda) \le t\Big\}\Big), \\
(\liminf |f_n|)^*(t) &\le \liminf(|f_n|^*(t)).
\end{aligned}$$

The estimate (iv) follows immediately from the latter inequality and Fatou's lemma. Properties (i), (ii), (v), and (vi) are left to the reader. □

Proposition 5.6. *Assume that W is a positive function on $(0,\infty)$. Let $\Lambda^p_{\mathfrak{X}}(w)$ be a Lorentz space and let $(f_n)_n$ be a sequence of measurable functions on $\mathfrak{X}$.*

(i) *If $\lim_{m,n} \|f_m - f_n\|_{\Lambda^p_{\mathfrak{X}}(w)} = 0$, then $(f_n)_n$ is a Cauchy sequence in measure and there exists $f \in \mathfrak{F}(\mathfrak{X}, \mathcal{A})$ such that $\lim_n \|f_n - f\|_{\Lambda^p_{\mathfrak{X}}(w)} = 0$.*

(ii) *If $f \in \mathfrak{F}(\mathfrak{X}, \mathcal{A})$ and $\lim_n \|f_n - f\|_{\Lambda^p_{\mathfrak{X}}(w)} = 0$, then $(f_n)_n$ converges to f in measure and there exists a subsequence $(f_{n_k})_k$ convergent to f a. e.*

Proof. The case $p = \infty$ is trivial as $\Lambda^p_{\mathfrak{X}}(w) = L_\infty$ and the result is already known. If $p < \infty$, it is immediate, by Proposition 5.3, that

$$W(D_f(t)) \leq \frac{\|f\|^p_{\Lambda^p_{\mathfrak{X}}(w)}}{t^p}, \qquad t > 0,$$

using the hypothesis of (*i*). We then obtain that $W(D_{f_m - f_n}(t)) \underset{m,n}{\longrightarrow} 0$ for every $t > 0$, which (since $W > 0$) implies $D_{f_m - f_n}(t) \underset{m,n}{\longrightarrow} 0$, $t > 0$, that is, $(f_n)_n$ is a Cauchy sequence in measure. We know that this implies the convergence in measure of $(f_n)_n$ to some measurable function f and the existence of a subsequence $(f_{n_k})_k$ converging to f a. e. By Proposition 5.5(iv), we have

$$\|f - f_n\|_{\Lambda^p_{\mathfrak{X}}(w)} \leq \liminf_k \|f_{n_k} - f_n\|_{\Lambda^p_{\mathfrak{X}}(w)},$$

and thus $\lim_n \|f - f_n\|_{\Lambda^p_{\mathfrak{X}}(w)} = 0$. The proof of (ii) is analogous. □

Remark 5.6. The functional $\|\cdot\|_{\Lambda^p_{\mathfrak{X}}}$ is not, in general, a quasinorm and, in fact, $\Lambda^p_{\mathfrak{X}}(w)$ is not even a vector space. Let us show this by means of the counterexample below.

Consider $(\mathbb{R}, \mathcal{L}, m)$. Set

$$w(t) = \mathcal{X}_{(0,1)}(t) + \frac{1}{1-t}\mathcal{X}_{(1,+\infty)}(t).$$

Taking $f(x) = \chi_{(0,1)}(x)$ and $g(x) = \chi_{(1,2)}(x)$, one has $f^*(t) = \chi_{(0,1)}(x)$ and $g^*(t) = \chi_{(0,1)}(t)$, and so

$$\begin{aligned}\int_0^{+\infty} [f^*(t)]^p w(t)\,dt &= \int_0^{+\infty} \mathcal{X}_{(0,1)}(t)\Big(\chi_{(0,1)}(t) + \frac{1}{1-t}\mathcal{X}_{(1,+\infty)}(t)\Big)\,dt \\ &= \int_0^{+\infty} \chi_{(0,1)}(t)\mathcal{X}_{(0,1)}(t)\,dt + \int_0^{+\infty} \frac{1}{1-t}\mathcal{X}_{(0,1)}(t)\,\mathcal{X}_{(1,+\infty)}(t)\,dt \\ &= 1,\end{aligned}$$

hence $f \in \Lambda^p_{\mathfrak{X}}(w)$. The same calculation shows $g \in \Lambda^p_{\mathfrak{X}}(w)$.

Note that

$$\begin{aligned}(f+g)^*(t) \leq f^*(t/2) + g^*(t/2) &= \mathcal{X}_{(0,1)}(t/2) + \mathcal{X}_{(0,1)}(t/2) \\ &= \mathcal{X}_{(0,2)}(t) + \mathcal{X}_{(0,2)}(t) \\ &= 2\,\mathcal{X}_{(0,2)}(t).\end{aligned}$$

Then

$$\begin{aligned}
\int_0^{+\infty} (f+g)^*(t)w(t)\,dt &= \int_0^{+\infty} 2\,\mathcal{X}_{(0,2)}(t)\,w(t)\,dt \\
&= 2\int_0^{+\infty} \mathcal{X}_{(0,2)}(t)\,\mathcal{X}_{(0,1)}(t)\,dt + 2\int_0^{+\infty} \frac{\mathcal{X}_{(0,2)}(t)\,\mathcal{X}_{(1,+\infty)}(t)}{1-t}\,dt \\
&= 2\int_0^{+\infty} \mathcal{X}_{(0,1)}(t)\,dt + 2\int_0^{+\infty} \frac{\mathcal{X}_{(0,2)\cap(1,+\infty)}(t)}{1-t}\,dt \\
&= 2\int_0^{1} dt + 2\int_0^{+\infty} \frac{\mathcal{X}_{(1,2)}(t)}{1-t}\,dt \\
&= 2 + 2\int_1^{2} \frac{1}{1-t}\,dt \\
&= 2 + 2\ln|1-t|\Big|_1^2 \to +\infty.
\end{aligned}$$

Therefore,

$$\|f+g\|_{\Lambda^p_{\mathfrak{X}}(w)} > \|f\|_{\Lambda^p_{\mathfrak{X}}(w)} + \|g\|_{\Lambda^p_{\mathfrak{X}}(w)},$$

and hence $f+g \notin \Lambda^p_{\mathfrak{X}}(w)$.

1. Observe that

$$\inf\{\lambda > 0 : D_f(\lambda) \le \mu(\mathfrak{X})\} = 0.$$

Hence, if $\mu(\mathfrak{X}) < +\infty$ and $t \ge \mu(\mathfrak{X}) \ge D_f(\lambda)$, if follows that

$$f^*(t) = \inf\{\lambda > 0 : D_f(\lambda) \le t\} = 0.$$

Therefore, the behavior of the weight w on $(\mu(\mathfrak{X}), +\infty)$ is irrelevant. This allows us to assume without loss of generality that the weight w vanishes on $(\mu(\mathfrak{X}), +\infty)$, if $\mu(\mathfrak{X}) < +\infty$.

If $\mu(\mathfrak{X}) = +\infty$, then $w \in L_1(\mathbb{R})$, since

$$+\infty > W(\mu(\mathfrak{X})) = \int_0^{\mu(\mathfrak{X})} w(t)\,dt = \int_0^{+\infty} w(t)\,dt.$$

2. Also, $0 < h \le r$ if and only if $W(h) \le W(r)$.
This comes from the fact that $W(h) = \int_0^h w(t)\,dt$ is a continuous and increasing function.

3. $W(Mr) \leq MW(r)$ if $M \geq 1$ and w is a nonincreasing weight. Indeed, observe that

$$W(Mr) = \int_0^{Mr} w(t)\,dt \overset{\boxed{t=pM}}{=} M\int_0^{r} w(pM)\,dp$$
$$\leq M\int_0^{r} w(p)\,dp$$
$$\leq MW(r).$$

Following an analogous procedure, one can verify that:
I. The inequality remains true if $M \leq 1$ and w is a nondecreasing weight.
II. The inequality is reversed if:
   i. $M \geq 1$ and w is a nondecreasing weight;
   ii. $M \leq 1$ and w is a nonincreasing weight.

(a) If $f \in \Lambda^p_{\mathfrak{X}}(w)$ and $w \notin L_1(\mathbb{R}_+)$, then

$$\lim_{t\to+\infty} f^*(t) = 0.$$

In fact, assume that $\lim_{t\to+\infty} f^*(t) = a > 0$. Then for a given $\epsilon > 0$, there exists $M > 0$ such that $a - \epsilon < f^*(t)$ if $t \geq M$. Then

$$\|f\|_{\Lambda^p_{\mathfrak{X}}(w)} = \left(\int_0^{+\infty} (f^*(t))^p w(t)\,dt\right)^{\frac{1}{p}}$$
$$\leq \left(\int_M^{+\infty} (a-\epsilon)^p w(t)\,dt\right)^{\frac{1}{p}}$$
$$= (a-\epsilon)\left(\int_M^{+\infty} w(t)\,dt\right)^{\frac{1}{p}}$$
$$= +\infty.$$

This yields a contradiction.

(b) All simple functions with finite support belong to $\Lambda^p_{\mathfrak{X}}(w)$. In fact, let s be a simple function and $\operatorname{supp}(s) = \overline{\{x \in \mathfrak{X} : s(x) \neq 0\}}$ with $\mu(\operatorname{supp}(s)) < +\infty$.

The following theorem characterizes the quasinormability of theses spaces, which, as we will see, only depends on the weight and on the measure space $\mathfrak{X}$.

Theorem 5.1. *If $0 < p < \infty$, the space $\Lambda^p_{\mathfrak{X}}(w)$ is quasinormed if and only if*

$$0 < W\big(\mu(A \cup B)\big) \le C\big(W\big(\mu(A)\big) + W\big(\mu(B)\big)\big) \tag{5.7}$$

for every pair of measurable sets $A, B \subset \mathfrak{X}$ with $\mu(A \cup B) > 0$.

Proof. (Sufficiency) The hypothesis implies that $W(\mu(A)) > 0$ if $\mu(A) > 0$. If $\|f\|_{\Lambda^p_{\mathfrak{X}}(w)} = 0$, by Proposition 5.1, we have $W(D_f(t)) = 0$, $t > 0$, and hence $D_f(t) = 0$ for every $t > 0$, that is, $f = 0$ a. e. It remains to show the quasi-triangular inequality and it is sufficient to prove it for positive functions. Let $0 \le f, g \in \Lambda^p_{\mathfrak{X}}(w)$ and $t > 0$. Then

$$\{f + g > t\} \subset \left\{f > \frac{t}{2}\right\} \cup \left\{g > \frac{t}{2}\right\}$$

and, by hypothesis,

$$0 < W(D_{f+g}(t)) \le C\left(W\left(D_f\left(\frac{t}{2}\right)\right) + W\left(D_g\left(\frac{t}{2}\right)\right)\right).$$

Since C does not depend on t, by Proposition 5.1, we have

$$\|f + g\|_{\Lambda^p_{\mathfrak{X}}(w)} \le C_p(\|f\|_{\Lambda^p_{\mathfrak{X}}(w)} + \|g\|_{\Lambda^p_{\mathfrak{X}}(w)}).$$

(Necessity) If A, B are two measurable sets with $\mu(A \cup B) > 0$, $\mathcal{X}_{A\cup B} \le \mathcal{X}_A + \mathcal{X}_B$ and, since $\Lambda^p_{\mathfrak{X}}(w)$ is quasinormed, we have

$$\begin{aligned}
0 < [W(\mu(A \cup B))]^{1/p} &= \|\mathcal{X}_{A\cup B}\|_{\Lambda^p_{\mathfrak{X}}(w)} \\
&\le \|\mathcal{X}_A + \mathcal{X}_B\|_{\Lambda^p_{\mathfrak{X}}(w)} \\
&\le C(\|\mathcal{X}_A\|_{\Lambda^p_{\mathfrak{X}}(w)} + \|\mathcal{X}_B\|_{\Lambda^p_{\mathfrak{X}}(w)}) \\
&= C_p([W(\mu(A))]^{1/p} + [W(\mu(B))]^{1/p}),
\end{aligned}$$

which is equivalent to the condition of the statement. □

Remark 5.7. Let w be a weight in $(0, \infty)$. We write $W \in \Delta_2(\mathfrak{X})$ if W satisfies (5.7). Therefore, Theorem 5.1 tells us that, for every $0 < p < \infty$, $\Lambda^p_{\mathfrak{X}}(w)$ is quasinormed if and only if $W \in \Delta_2(\mathfrak{X})$. Also, the following conditions are equivalent:

- $W \in \Delta_2(\mathfrak{X})$;
- $W(2r) \le CW(r)$, $r > 0$;
- $W(t + s) \le C(W(t) + W(s))$, $t, s > 0$,

and in any of these cases, $W(t) > 0$, $t > 0$.

Observe that all of these properties are independent of the measure space $\mathfrak{X}$.

Theorem 5.2. *The space $(\Lambda_X^p(w), \|\cdot\|_{\Lambda_X^p(w)})$ is a Banach space.*

For more details on Lambda spaces, see [20].

5.3 Gamma spaces

Let f^{**} denote the maximal function of f^*, defined by

$$f^{**}(t) = \frac{1}{t}\int_0^t f^*(s)ds, \quad t > 0.$$

Some elementary properties of f^{**} are:

1. $f^* \leq f^{**}$;
2. f^{**} is nonincreasing;
3. $(f+g)^{**} \leq f^{**} + g^{**}$.

Definition 5.4. Let $\mathcal{F}(X,\mathcal{A})$ be the linear space of all $\mathcal{A}$-measurable functions on X and $1 \leq p < \infty$. Define the Gamma space $\Gamma_X^p(w)$ to be the set of all functions $f \in \mathcal{F}(X,\mathcal{A})$ such that

$$\|f\|_{\Gamma_X^p(w)} = \left(\int_0^\infty [f^{**}(t)]^p w(t)dt\right)^{\frac{1}{p}} < \infty.$$

In other words, the Gamma space is

$$\Gamma_X^p(w) = \left\{f \in \mathcal{F}(X,\mathcal{A}) : \|f\|_{\Gamma_X^p(w)} < \infty\right\}.$$

Observe that

$$\|f\|_{\Gamma_X^p(w)} = \left(\int_0^\infty [f^{**}(t)]^p w(t)dt\right)^{\frac{1}{p}} = \|f^{**}\|_{L_p(w)}. \tag{5.8}$$

Equation (5.8) tells us that

$$f \in \Gamma_X^p(w) \quad \text{if and only if} \quad f^{**} \in L_p(w).$$

The following denotes the weak Gamma spaces:

$$\Gamma_X^{p,\infty}(w) = \left\{f \in \mathcal{F}(X,\mathcal{A}) : \|f\|_{\Gamma_X^{p,\infty}(w)} = \sup_{t>0} f^{**}(t)[W(t)]^{\frac{1}{p}} < \infty\right\},$$

where

$$W(r) = \int_0^r w(t)dt.$$

We claim that $\Gamma_X^p(w)$ is a vector space. Indeed, for $f^{**}, g^{**} \in L_p(w)$ and from the fact that

$$(f+g)^{**}(t) \leq f^{**}(t) + g^{**}(t),$$

we have

$$\begin{aligned}
\left(\int_0^\infty [(f+g)^{**}(t)]^p w(t)dt\right)^{\frac{p}{q}} &\leq \int_0^\infty (f^{**}(t) + g^{**}(t))^p w(t)dt \\
&\leq \int_0^\infty (2\max\{|f^{**}(t)|, |g^{**}(t)|\})^p w(t)dt \\
&= 2^p \int_0^\infty (\max\{|f^{**}(t)|, |g^{**}(t)|\})^p w(t)dt \\
&\leq 2^p \left[\int_0^\infty |f^{**}(t)|^p w(t)dt + \int_0^\infty |g^{**}(t)|^p w(t)dt\right] \\
&< \infty.
\end{aligned}$$

Hence $(f+g)^{**} \in L_p(w)$, and so $f+g \in \Gamma_X^p(w)$.

Theorem 5.3. *If $1 < p < \infty$, then:*
(a) $\|f\|_{\Gamma_X^p(w)} = (\int_0^\infty |f^{**}(t)|^p w(t)dt)^{\frac{1}{p}}$ *is a norm on* $\Gamma_X^p(w)$;
(b) $\|f\|_{\Gamma_X^{p,\infty}(w)} = \sup_{t>0} f^{**}(t)[W(t)]^{\frac{1}{p}}$ *is a norm on* $\Gamma_X^{p,\infty}(w)$.

Proof. (a) Since $f^{**} \geq 0$ and w is positive, one has $\|f\|_{\Gamma_X^p(w)} \geq 0$.
For $\alpha \in \mathbb{R}$, $(\alpha f^{**})(t) = |\alpha| f^{**}(t)$, thus

$$\begin{aligned}
\|\alpha f\|_{\Gamma_X^p(w)} &= \left(\int_0^\infty [(\alpha f)^{**}(t)]^p w(t)dt\right)^{\frac{1}{p}} \\
&= \left(\int_0^\infty [|\alpha| f^{**}(t)]^p w(t)dt\right)^{\frac{1}{p}} \\
&= |\alpha| \left(\int_0^\infty [f^{**}(t)]^p w(t)dt\right)^{\frac{1}{p}} \\
&= |\alpha| \|f\|_{\Gamma_X^p(w)}.
\end{aligned}$$

Since $f^{**} = 0$ implies $f = 0$, we obtain

$$\begin{aligned}
\|f\|_{\Gamma_X^p(w)} = 0 &\Longleftrightarrow \left(\int_0^\infty [f^{**}(t)]^p w(t)dt\right)^{\frac{1}{p}} = 0 \\
&\Longleftrightarrow \int_0^\infty [f^{**}(t)]^p w(t)dt = 0 \\
&\Longleftrightarrow [f^{**}(t)]^p w(t)dt = 0 \quad \text{a.e.} \\
&\Longleftrightarrow [f^{**}(t)]^p = 0 \quad \text{a.e.} \\
&\Longleftrightarrow f^{**} = 0 \quad \text{a.e.} \\
&\Longleftrightarrow f = 0 \quad \text{a.e.}
\end{aligned}$$

Finally, by (5.8), $(f+g)^{**} \in L_p(w)$.

Now, note that

$$\begin{aligned}
\|[(f+g)^{**}]^{p-1}\|_{L_q(w)} &= \left(\int_0^\infty [(f+g)^{**}(t)]^{q(p-1)} w(t)dt\right)^{\frac{1}{q}} \\
&= \left(\left(\int_0^\infty [(f+g)^{**}(t)]^p w(t)dt\right)^{\frac{1}{p}}\right)^{\frac{p}{q}} \\
&= \|[(f+g)^{**}]\|_{L_p(w)}^{\frac{p}{q}} < \infty.
\end{aligned}$$

Therefore $[(f+g)^{**}]^{p-1} \in L_q(w)$.

Next, if $f^{**}, g^{**} \in L_q(w)$ and $\frac{1}{p} + \frac{1}{q} = 1$, by Hölder's inequality, we have

$$\begin{aligned}
&\int_0^\infty [(f+g)^{**}(t)]^p w(t)dt \\
&\quad = \int_0^\infty [(f+g)^{**}(t)][(f+g)^{**}(t)]^{p-1} w(t)dt \\
&\quad \le \int_0^\infty f^{**}(t)[(f+g)^{**}(t)]^{p-1} w(t)dt + \int_0^\infty g^{**}(t)[(f+g)^{**}(t)]^{p-1} w(t)dt \\
&\quad \le \left[\left(\int_0^\infty [f^{**}(t)]^p w(t)dt\right)^{\frac{1}{p}} + \left(\int_0^\infty [g^{**}(t)]^p w(t)dt\right)^{\frac{1}{p}}\right] \\
&\qquad \times \left(\int_0^\infty [(f+g)^{**}(t)]^p w(t)dt\right)^{\frac{1}{q}}.
\end{aligned}$$

Then

$$\left(\int_0^\infty [(f+g)^{**}(t)]^p w(t)dt\right)^{\frac{1}{p}} \le \left(\int_0^\infty [f^{**}(t)]^p w(t)dt\right)^{\frac{1}{p}} + \left(\int_0^\infty [g^{**}(t)]^p w(t)dt\right)^{\frac{1}{p}},$$

and so

$$\|f+g\|_{\Gamma_X^p(w)} \le \|f\|_{\Gamma_X^p(w)} + \|g\|_{\Gamma_X^p(w)},$$

completing the proof of part (a).

(b) Since $W(t) > 0$ and $f^{**} = 0 \Rightarrow f = 0$ a. e., we get

$$\begin{aligned}\|f\|_{\Gamma_X^{p,\infty}(w)} = 0 &\Longleftrightarrow \sup_{t>0} f^{**}(t)[W(t)]^{\frac{1}{p}} = 0 \\ &\Longleftrightarrow f^{**}(t)[W(t)]^{\frac{1}{p}} = 0, \quad \forall t \\ &\Longleftrightarrow f^{**} = 0 \\ &\Longleftrightarrow f = 0 \quad \text{a. e.}\end{aligned}$$

Since, $W(r) = \int_0^r w(t)dt$ and f^{**} are nonnegative, we have

$$\|f\|_{\Gamma_X^{p,\infty}(w)} = \sup_{t>0} f^{**}(t)[W(t)]^{\frac{1}{p}} \ge 0.$$

Furthermore,

$$\begin{aligned}\|\alpha f\|_{\Gamma_X^{p,\infty}(w)} &= \sup_{t>0} (\alpha f)^{**}(t)[W(t)]^{\frac{1}{p}} \\ &= \sup_{t>0} |\alpha| f^{**}(t)[W(t)]^{\frac{1}{p}} \\ &= |\alpha| \sup_{t>0} f^{**}(t)[W(t)]^{\frac{1}{p}} \\ &= |\alpha| \|f\|_{\Gamma_X^{p,\infty}(w)}.\end{aligned}$$

Finally,

$$\begin{aligned}\|f+g\|_{\Gamma_X^{p,\infty}(w)} &= \sup_{t>0} (f+g)^{**}(t)[W(t)]^{\frac{1}{p}} \\ &\le \sup_{t>0} [f^{**}(t) + g^{**}(t)][W(t)]^{\frac{1}{p}}\end{aligned}$$

$$
\begin{aligned}
&= \sup_{t>0}\left(f^{**}(t)[W(t)]^{\frac{1}{p}} + g^{**}(t)[W(t)]^{\frac{1}{p}}\right) \\
&\le \sup_{t>0} f^{**}(t)[W(t)]^{\frac{1}{p}} + \sup_{t>0} g^{**}(t)[W(t)]^{\frac{1}{p}} \\
&\le \|f\|_{\Gamma_X^{p,\infty}(w)} + \|g\|_{\Gamma_X^{p,\infty}(w)},
\end{aligned}
$$

finishing the proof. □

Corollary 5.1. *The spaces* $(\Gamma_X^p(w), \|\cdot\|_{\Gamma_X^p(w)})$ *and* $(\Gamma_X^{p,\infty}(w), \|\cdot\|_{\Gamma_X^{p,\infty}(w)})$ *are normed vector spaces.*

Before moving on, we want to state and prove a simple, but no less important result.

Proposition 5.7. *Let w be a nonincreasing weight.*
(i) *If $b \ge 1$, then $w(bx) \le w(x)$.*
(ii) *If $b \ge 1$, then $W(bx) \le bW(x)$.*
(iii) *If $b \le 1$, then $w(x) \le w(bx)$.*
(iv) *If $b \le 1$, then $bW(x) \le W(bx)$.*

Proof. (i) Let $b \ge 1$, then $bx \ge x$ for $x > 0$ and, by the hypothesis on w, we have

$$w(bx) \le w(x).$$

(ii)

$$
\begin{aligned}
W(bx) &= \int_0^{bx} w(s)ds \quad (bt = s) \\
&= \int_0^{x} w(bt)bdt \\
&\le b\int_0^{x} w(t)dt \\
&= bW(x),
\end{aligned}
$$

where we have used (i).

Claims (iii) and (iv) are proved analogously, and the proof is complete. □

Proposition 5.8. *Let $f \in \mathcal{F}(X, \mathcal{A})$ and w be a decreasing weight. Then*
(i) $\int_0^\infty [f^{**}(\frac{t}{a})]^p w(t)dt \le a \int_0^\infty [f^{**}(t)]^p w(t)dt$, *for $a \ge 1$;*
(ii) $\int_0^\infty [f^{**}(\frac{t}{a})]^p w(t)dt \ge a \int_0^\infty [f^{**}(t)]^p w(t)dt$, *for $a \le 1$.*

Proof. (i) For $a \geq 1$, we have

$$\int_0^\infty \left[f^{**}\left(\frac{t}{a}\right)\right]^p w(t)dt = \int_0^\infty [f^{**}(u)]^p w(au)adu \quad \left(u = \frac{t}{a}\right)$$
$$\leq a \int_0^\infty [f^{**}(u)]^p w(u)du.$$

The proof of (ii) is similar to that of (i), and we are done. □

Theorem 5.4. *Let $(X, \mathcal{A}, \mu)$ be a σ-finite measure space. If w is a nonincreasing weight, then the norms $\|\cdot\|_{\Gamma_X^p(w)}$ and $\|\cdot\|_{\Lambda_X^p(w)}$ are equivalent.*

Proof. Consider $f \in \mathcal{F}(X, \mathcal{A})$. On the one hand, invoking Theorem [19, Theorem 2.6], and since w is nonincreasing, we have

$$\|f\|_{\Gamma_X^p(w)} = \left(\int_0^\infty [f^{**}(x)]^p w(x)dx\right)^{\frac{1}{p}}$$
$$= \left(\int_0^\infty \left[\frac{1}{x}\int_0^x f^*(t)dt\right]^p w(x)dx\right)^{\frac{1}{p}}$$
$$\leq \frac{p}{p-1}\left(\int_0^\infty [f^*(x)]^p w(x)dx\right)^{\frac{1}{p}}$$
$$= \frac{p}{p-1}\|f\|_{\Lambda_X^p(w)}.$$

On the other hand, since $f^* \leq f^{**}$, we have $\|f\|_{\Lambda_X^p(w)} \leq \|f\|_{\Gamma_X^p(w)}$.

Finally, we conclude that

$$\|f\|_{\Lambda_X^p(w)} \leq \|f\|_{\Gamma_X^p(w)} \leq \frac{p}{p-1}\|f\|_{\Lambda_X^p(w)},$$

as claimed. □

Theorem 5.5. *The space $(\Gamma_X^p(w), \|\cdot\|_{\Gamma_X^p(w)})$ is a Banach space.*

Proof. Simply apply Theorems 5.2 and 5.4. □

Lemma 5.4. *Let $p \in (1, \infty)$. Then there exists a constant C such that for all $a_k \geq 0$,*

$$\sum_{j=-\infty}^{\infty} 2^j \left(\sum_{k=j}^{\infty} a_k \right)^p \leq C \left(\sum_{j=-\infty}^{\infty} 2^k a_k^p \right) \quad \text{holds.}$$

Proof. By Hölder's inequality for sums, we obtain

$$\begin{aligned}
\sum_{j=-\infty}^{\infty} 2^j \left(\sum_{k=j}^{\infty} a_k \right)^p &= \sum_{j=-\infty}^{\infty} 2^j \left(\sum_{k=j}^{\infty} 2^{\frac{j-k}{pq}} 2^{\frac{k-j}{pq}} a_k \right)^p \\
&\leq \sum_{j=-\infty}^{\infty} 2^j \left(\sum_{k=j}^{\infty} 2^{\frac{j-k}{p}} \right)^{p-1} \left(\sum_{k=j}^{\infty} 2^{\frac{k-j}{q}} a_k^p \right) \\
&\leq C \left[\sum_{j=-\infty}^{\infty} 2^j \left(\sum_{k=j}^{\infty} 2^{\frac{k-j}{q}} a_k^p \right) \right] \\
&= C \sum_{j=-\infty}^{\infty} 2^{\frac{k}{q}} a_k^p \sum_{j=-\infty}^{k} 2^{\frac{j}{p}} \\
&\leq C \sum_{j=-\infty}^{\infty} 2^k a_k^p.
\end{aligned}$$

□

Theorem 5.6. *Let $1 < p < \infty$, g, v be nonnegative measurable functions, and let $v \in L^1_{\mathrm{loc}}([0, \infty))$. Denote $G(t) = \int_0^t g(s)ds$ and $V(t) = \int_0^t v(s)ds$. Then*

$$\begin{aligned}
\sup_{\substack{f\downarrow \\ f \geq 0}} \frac{\int_0^\infty f(x)g(x)dx}{(\int_{-\infty}^a (f(x))^p v(x)dx)^{\frac{1}{p}}} &\approx \int_0^\infty \left(\int_x^\infty \frac{g(t)}{V(t)} dt \right)^q v(x)dx \\
&\approx \left(\int_0^\infty (G(x))^{q-1} (V(x))^{1-q} g(x)dx \right)^{\frac{1}{q}} \\
&\approx \left(\int_0^\infty (G(x))^q \frac{v(x)}{(V(x))^q} dx \right)^{\frac{1}{q}} + \frac{\int_0^\infty g(s)ds}{(\int_0^\infty v(s)ds)^{\frac{1}{p}}},
\end{aligned} \tag{5.9}$$

where the constant in $\approx$ depends only on p.

Proof. Due to the monotone convergence theorem, we may assume that g has compact support (for g without compact support, consider $g_n = \chi_{(0,n)} g$ and if for all g_n the inequality holds, then by the monotone convergence theorem it holds for g as well). Also, without loss of generality, suppose $\int_0^\infty g(s)ds = 1$ (otherwise, we can take ag with an appropriate $a \in (0, \infty)$). Define $V(t) = \int_0^t v(s)ds$. Then $1 < V(t) < \infty$, for all $t > 0$. Set

$$\varphi(x) = \left(\int_x^\infty \frac{g(s)}{V(s)} ds \right)^{q-1}$$

for $x \in (0, \infty)$. Then φ is bounded and nonincreasing on $(0, \infty)$. Integration by parts yields

$$\begin{aligned}\int_0^\infty (\varphi(x))^p v(x) dx &= \left[V(x) \left(\int_x^\infty \frac{g(t)}{V(t)} dt \right)^q dt \right]_0^\infty + q \int_0^\infty g(x) \left(\int_x^\infty \frac{g(t)}{V(t)} dt \right)^{q-1} dx \\ &= q \int_0^\infty g(x)\varphi(x) dx.\end{aligned}$$

The expression $[V(x) \int_x^\infty \frac{g(t)}{V(t)} dt]_0^\infty$ equals zero since $\operatorname{supp}(g) \subset (0, \infty)$ is compact. Therefore taking $f(x) = \varphi(x)$, we have that the supremum on the left-hand side is at least

$$\begin{aligned}\frac{\int_0^\infty \varphi(s) g(s) ds}{(\int_0^\infty (\varphi(s))^p v(s) ds)^{\frac{1}{p}}} &= \frac{\int_0^\infty (\varphi(s))^p v(s) ds}{q(\int_0^\infty (\varphi(s))^p v(s) ds)^{\frac{1}{p}}} \\ &= \frac{1}{q} \left(\int_0^\infty (\varphi(s))^p v(s) ds \right)^{\frac{1}{q}} \\ &= \frac{1}{q} \left(\left(\int_x^\infty \frac{g(t)}{V(t)} ds \right)^q v(t) dt \right)^{\frac{1}{p}}.\end{aligned}$$

Now, considering f to be a nonnegative and nonincreasing function, by Fubini's theorem we have

$$\begin{aligned}\int_0^\infty f(s) g(s) ds &= \int_0^\infty f(s) \frac{g(s)}{V(s)} \int_0^\infty v(t) dt ds \\ &= \int_0^\infty \left(\int_t^\infty \frac{f(s) g(s)}{V(s)} ds \right) v(t) dt \\ &\leq \int_0^\infty f(t) \left(\int_t^\infty \frac{g(s)}{V(s)} ds \right) v(t) dt \\ &\leq \left(\int_t^\infty [f(t)]^p v(t) dt \right)^{\frac{1}{p}} \left(\int_0^\infty \left(\int_t^\infty \frac{g(s)}{V(s)} ds \right)^q v(t) dt \right)^{\frac{1}{q}}.\end{aligned}$$

The last inequality follows from Hölder's inequality and the second to last comes from the fact that f is nonincreasing. So we are done with the equivalence of the left-hand side and the first expression on the right of equation (5.9).

On the one hand, let $\{x_j\}_{j\in\mathbb{N}}$ be a sequence satisfying $\int_0^{x_j} g(s)ds = 2^{-j}$. Then

$$
\begin{aligned}
\int_0^\infty\left(\int_t^\infty \frac{g(s)}{V(s)}ds\right)^{q-1} g(t)dt &= \sum_{j=0}^\infty \int_{x_{j+1}}^{x_j}\left(\int_t^\infty \frac{g(s)}{V(s)}ds\right)^{q-1} g(t)dt \\
&\ge \sum_{j=0}^\infty \int_{x_{j+1}}^{x_j}\left(\int_t^\infty \frac{g(s)}{V(x_j)}ds\right)^{q-1} g(t)dt \\
&= \sum_{j=0}^\infty (V(x_j))^{1-q}\left(\int_{x_j}^\infty g(s)ds\right)^{q-1} \int_{x_{j+1}}^{x_j} g(t)dt \\
&\ge C\sum_{j=0}^\infty (V(x_j))^{1-q}\left(\int_{x_j}^\infty g(s)ds\right)^{q-1} \int_{x_{j+1}}^{x_{j-1}} g(t)dt \\
&= C\sum_{j=0}^\infty (V(x_{j+1}))^{1-q}\left(\int_{x_{j+1}}^\infty g(s)ds\right)^{q-1} \int_{x_{j+1}}^{x_{j-1}} g(t)dt \\
&\ge C\sum_{j=0}^\infty \int_{x_{j+1}}^{x_j} (G(x_j))^{q-1}(V(x_{j+1}))^{1-q} g(t)dt \\
&\ge C\int_0^\infty (G(t))^{q-1}(V(t))^{1-q} g(t)dt.
\end{aligned}
$$

On the other hand, if $\int_0^\infty v(t)dt = \infty$, set $N = \infty$.

In the case of $\int_0^\infty v(t)dt < \infty$, let N denote the largest integer for which $2^{N-1} < \int_0^\infty v(t)dt$. Then

$$
\begin{aligned}
\int_0^\infty\left(\int_t^\infty \frac{g(s)}{V(s)}ds\right)^{q} v(t)dt &= \sum_{j=-\infty}^{N-1} \int_{x_j}^{x_{j+1}}\left(\int_t^\infty \frac{g(s)}{V(s)}ds\right)^{q} v(t)dt \\
&= \sum_{j=-\infty}^{N-1}\left(\int_{x_j}^{x_{j+1}} v(t)dt\right)\left(\int_t^\infty \frac{g(s)}{V(s)}ds\right)^{q} \\
&\le \sum_{j=-\infty}^{N-1}\left(\int_{x_j}^{x_{j+1}} v(t)dt\right)\left(\sum_{k=j}^{N-1} \frac{\int_{x_k}^{x_{k+1}} g(s)ds}{V(x_k)}\right)^{q} \\
&\le C\sum_{j=-\infty}^{N-1} 2^j\left(\sum_{k=j}^{N-1} 2^{-k}\int_{x_k}^{x_{k+1}} g(s)ds\right)^{q}
\end{aligned}
$$

$$\leq C \sum_{j=-\infty}^{N-1} 2^j \left(2^{-j} \int_{x_j}^{x_{j+1}} g(s)ds \right)^q$$

$$= C \sum_{j=-\infty}^{N-1} \left(\frac{V(x_{j+1})}{2} \right)^{q-1} \int_{x_j}^{x_{j+1}} q \left(\int_{x_j}^{t} g(s)ds \right)^{q-1} g(t)dt$$

$$\leq C \sum_{j=-\infty}^{N-1} \int_{x_j}^{x_{j+1}} (G(t))^{q-1} (V(t))^{1-q} g(t)dt$$

$$= C \int_0^\infty (G(t))^{q-1} (V(t))^{1-q} g(t)dt,$$

where the inequality between the fourth and fifth line follows from Lemma 5.4, applied to the sequence

$$a_k = 2^{-k} \int_{x_k}^{x_{k+1}} g(s)ds \quad \text{for } -\infty \leq K < N,$$

$$a_k = 0 \quad \text{for } K \geq N.$$

To complete the proof, let us integrate the last expression by parts:

$$\int_0^\infty (G(t))^{q-1} (V(t))^{1-q} g(t)dt$$

$$= \frac{1}{q} [(G(t))^q (V(t))^{1-q}]_0^\infty + \frac{1}{p} \int_0^\infty (G(t))^q (V(t))^{-q} v(t)dt$$

$$= \frac{1}{p} \int_0^\infty (G(t))^q \frac{v(t)}{(V(t))^q} dt + \frac{1}{q} \left(\int_0^\infty g(s)ds \right)^q \left(\int_0^\infty v(s)ds \right)^{1-q},$$

which proves the equivalence between the second and third expression on the right-hand side of (5.9). The proof is complete. □

Corollary 5.2. *Let $v \in L^1_{\text{loc}}$, $v > 0$ be a nonincreasing function and $\int_0^\infty v(t)dt = \infty$. Set $X = \Lambda^p_X(w)$. Then $\|g\|_{X'} \approx \|g\|_{\Gamma^q_X(w)}$, where $w(x) = x^q \frac{v(x)}{[V(x)]^q}$.*

Proof. By Theorem 5.2, $\Lambda^p_X(w)$ is a Banach function space. Therefore we may consider its associate space with the norm defined by

$$\|g\|_{X'} = \sup_{\|f\|_{\Lambda_X^p(w)} \le 1} \int_X fg d\mu$$
$$= \sup_{f>0} \frac{\int_X fg d\mu}{(\int_0^\infty (f^*(s))^p v(t) ds)^{\frac{1}{p}}},$$

where we use the convention $\frac{0}{0} = 0$. Since (X, μ) is nonatomic and therefore resonant, we can replace $\int_X fg d\mu$ by $\int_0^\infty f^*(t) g^*(t) dt$.

Hence

$$\|g\|_{X'} = \sup_{f \ge 0} \frac{\int_0^\infty f^*(t) g^*(t) dt}{([f^*(t)]^p v(t) dt)^{\frac{1}{p}}},$$

which, by Theorem 5.6, is equivalent to

$$\left(\int_0^\infty [G(x)]^q \frac{v(x)}{[V(x)]^q} dx \right)^{\frac{1}{q}} + \frac{\int_0^\infty g(t) dt}{(\int_0^\infty v(t) dt)^{\frac{1}{p}}}$$

and, since $\int_0^\infty v(t) dt = \infty$, the latter expression in the sum equals zero. Hence, we have

$$\left(\int_0^\infty [G(x)]^q \frac{v(x)}{[V(x)]^q} dx \right)^{\frac{1}{q}} = \left(\int_0^\infty [g^{**}(x)]^q x^q \frac{v(x)}{[V(x)]^q} dx \right)^{\frac{1}{q}}$$
$$= \|g\|_{\Gamma_X^q(w)},$$

completing the proof. □

5.4 Weighted Lorentz spaces

In the sequel, by taking the measure $\omega d\mu$ instead of μ, we shall consider the measure space $(X, \omega d\mu)$ and define on it the weighted distribution function as follows:

$$D_f^\omega(\lambda) = \omega(\{x \in X : |f(x)| > \lambda\})$$
$$= \int_{\{x \in X : |f(x)| > \lambda\}} \omega(x) \, d\mu(x), \quad \lambda \ge 0.$$

The nonnegative weighted rearrangement of f is given by

$$f_\omega^*(t) = \inf\{\lambda > 0 : D_f^\omega(\lambda) \le t\}$$
$$= \sup\{\lambda > 0 : D_f^\omega(\lambda) \ge t\}, \quad t \ge 0,$$

where we assume that $\inf \emptyset = \infty$ and $\sup \emptyset = 0$.

Also, the average (maximal) function of f on $(0,\infty)$ is given by

$$f_\omega^{**}(t) = \frac{1}{t}\int_0^t f_\omega^*(s)\,ds.$$

Proposition 5.9. *We have the following properties:*
(a) $D_f^\omega(\cdot)$ *is decreasing and continuous from the right;*
(b) $f_\omega^*(\cdot)$ *is decreasing;*
(c) $f_\omega^*(\cdot) \le f_\omega^{**}(\cdot)$*;*
(d) $f_\omega^{**}(\cdot)$ *is decreasing.*

Proof. (a) Let $0 \le \lambda_1 \le \lambda_2$ be arbitrary. Then

$$\{x \in X : |f(x)| > \lambda_2\} \subset \{x \in X : |f(x)| > \lambda_1\}.$$

By the monotonicity of the integral, we have

$$\begin{aligned} D_f^{(\omega)}(\lambda_2) &= \omega(\{x \in X : |f(x)| > \lambda_2\}) \\ &= \int_{\{x\in X:|f(x)|>\lambda_2\}} \omega(x)\,d\mu(x) \\ &\le \int_{\{x\in X:|f(x)|>\lambda_1\}} \omega(x)\,d\mu(x) \\ &= D_f^{(\omega)}(\lambda_1), \end{aligned}$$

and thus

$$D_f^{(\omega)}(\lambda_2) \le D_f^{(\omega)}(\lambda_1).$$

Let $\lambda_0 \ge 0$ and define

$$E_f(\lambda) = \{x \in X : |f(x)| > \lambda\}.$$

Notice that

$$E_f(\lambda_0) = \bigcup_{n=1}^{\infty} E_f\left(\lambda_0 + \frac{1}{n}\right).$$

Hence

$$\begin{aligned}\lim_{n\to\infty} D_f^{\omega}\left(\lambda_0+\frac{1}{n}\right) &= \lim_{n\to\infty} \omega\left(\left\{x\in X : |f(x)| > \lambda_0+\frac{1}{n}\right\}\right)\\ &= \lim \omega\left(E_f\left(\lambda_0+\frac{1}{n}\right)\right) = \lim_{n\to\infty} \int\limits_{E_f(\lambda_0+\frac{1}{n})} \omega(x)\,d\mu(x)\\ &= \int\limits_{\bigcup_{n=1}^{\infty} E_f(\lambda_0+\frac{1}{n})} \omega(x)\,d\mu(x) = \int\limits_{E_f(\lambda_0)} \omega(x)\,d\mu(x)\\ &= \omega(E_f(\lambda_0)) = \omega(\{x\in X : |f(x)| > \lambda_0\})\\ &= D_f^{\omega}(\lambda_0).\end{aligned}$$

Therefore,

$$\lim_{n\to\infty} D_f^{\omega}\left(\lambda_0+\frac{1}{n}\right) = D_f^{\omega}(\lambda_0).$$

(b) Let $0 \le t \le u$. Since

$$\{\lambda \ge 0 : D_f^{\omega}(\lambda) \le t\} \subset \{\lambda \ge 0 : D_f^{\omega}(\lambda) \le u\},$$

we get

$$\inf\{\lambda \ge 0 : D_f^{\omega}(\lambda) \le u\} \le \inf\{\lambda \ge 0 : D_f^{\omega}(\lambda) \le t\},$$

and thus

$$f_{\omega}^{*}(u) \le f_{\omega}^{*}(t).$$

(c) Since $f_{\omega}^{*}(\cdot)$ is decreasing,

$$\begin{aligned}f_{\omega}^{**}(t) &= \frac{1}{t}\int_0^t f_{\omega}^{*}(s)\,ds \ge \int_0^t f_{\omega}^{*}(t)\,ds\\ &= f_{\omega}^{*}(t),\end{aligned}$$

and hence

$$f_{\omega}^{*}(t) \le f_{\omega}^{**}(t).$$

(d) If $0 < t \le s$, due to the fact that f^{*} is decreasing, we have $f^{*}(v) \le f^{*}(\frac{tv}{s})$, hence

$$f_\omega^{**}(s) = \frac{1}{s}\int_0^s f_\omega^*(v)\,dv \le \frac{1}{s}\int_0^s f_\omega^*\left(\frac{tv}{s}\right)dv$$
$$= \frac{1}{t}\int_0^t f_\omega^*(u)\,du = f_\omega^{**}(t),$$

that is,

$$f_\omega^{**}(s) \le f_\omega^{**}(t),$$

and so $f_\omega^{**}(\cdot)$ is decreasing. □

Definition 5.5. The weighted Lorentz space $L(p,q,\omega d\mu)$ is the collection of all functions f such that

$$\|f\|_{p,q,\omega} = \begin{cases} (\frac{q}{p}\int_a^b t^{\frac{q}{p}-1}[f_\omega^*(t)]^q\,dt)^{1/q}, & 0 < p, q < \infty; \\ \sup_{t>0} t^{\frac{1}{p}} f_\omega^*(t), & 0 < p < q = \infty. \end{cases}$$

In general, $\|\cdot\|_{p,q,\omega}$ is not a norm since Minkowski inequality may fail. But by replacing f_ω^* with f_ω^{**} in the above definition of $\|\cdot\|_{(p,q),\omega}$, we obtain a norm $\|\cdot\|_{(p,q),\omega}$ for all $q \ge 1$ given by

$$\|f\|_{(p,q),\omega} = \begin{cases} (\frac{q}{p}\int_a^b t^{\frac{q}{p}-1}[f_\omega^{**}(t)]^q\,dt)^{1/q}, & 1 < p < \infty,\ 0 < \infty \le \infty; \\ \sup_{t>0} t^{\frac{1}{p}} f_\omega^{**}(t), & 1 < p \le \infty,\ q = \infty. \end{cases}$$

We shall denote by $L((p,q),\omega d\mu)$ the collection of all functions f such that $\|f\|_{(p,q),\omega} < \infty$.

Example 5.1. Let $E \in \mathcal{A}$ with $\omega(E) < \infty$, and consider $f = \chi_E$, then

$$\{x \in X : \chi_E(x) > \lambda\} = \begin{cases} E & \text{if } 0 \le \lambda < 1, \\ \emptyset & \text{if } \lambda \ge 1. \end{cases}$$

Hence

$$\omega(\{x \in X : \chi_E(x) > \lambda\}) = \begin{cases} \omega(E) & \text{if } 0 \le \lambda < 1, \\ 0 & \text{if } \lambda \ge 1, \end{cases}$$

and thus

$$D_f^\omega(\lambda) = \begin{cases} \omega(E) & \text{if } 0 \le \lambda < 1, \\ 0 & \text{if } \lambda \ge 1. \end{cases}$$

Therefore

$$f_\omega^*(t) = \inf\{\lambda \ge 0 : D_f^\omega(\lambda) \le t\} = \begin{cases} 1 & \text{if } t < \omega(E), \\ 0 & \text{if } t \ge \omega(E), \end{cases}$$

which means that

$$f_\omega^*(t) = \chi_{[0,\omega(E))}(t).$$

Next,

$$\begin{aligned}(\chi_E)_\omega^{**}(t) &= \frac{1}{t}\int_0^t (\chi_E)_\omega^*(s)\,ds \\ &= \frac{1}{t}\int_0^t \chi_{[0,\omega(E))}(t)\,ds \\ &= \begin{cases} 1 & \text{if } 0 \le t \le \omega(E), \\ \frac{\omega(E)}{t} & \text{if } t \ge \omega(E) \end{cases} \\ &= \min\left(1, \frac{\omega(E)}{t}\right).\end{aligned}$$

Now, for $1 \le q < \infty$,

$$\begin{aligned}\|\chi_E\|_{(p,q),\omega}^q &= \frac{q}{p}\int_0^\infty \left(t^{\frac{1}{p}}(\chi_E)_\omega^{**}(t)\right)^q \frac{dt}{t} \\ &= \frac{q}{p}\left[\int_0^{\omega(E)} t^{\frac{q}{p}-1}\,dt + \int_{\omega(E)}^\infty \left(t^{\frac{1}{p}}\frac{\omega(E)}{t}\right)^q \frac{dt}{t}\right] \\ &= \frac{q}{p}\left[\int_0^{\omega(E)} t^{\frac{q}{p}-1}\,dt + [\omega(E)]^q \int_{\omega(E)}^\infty t^{\frac{q}{p}-q-1}\,dt\right] \\ &= \frac{q}{p}\left[\frac{p}{q}[\omega(E)]^{\frac{q}{p}} + \frac{t^{\frac{q}{p}-q}}{\frac{q}{p}-q}\Bigg|_{\omega(E)}^\infty [\omega(E)]^q\right].\end{aligned}$$

Note that, since $1 < p$, one has $1 - p < 0$, and so $\frac{1}{p} - 1 < 0$. Hence $\frac{q}{p} - q < 0$ and

$$\begin{aligned}\frac{q}{p}\left[\frac{p}{q}[\omega(E)]^{\frac{q}{p}} + \frac{t^{\frac{q}{p}-q}}{\frac{q}{p}-q}\Bigg|_{\omega(E)}^\infty [\omega(E)]^q\right] &= \frac{q}{p}\left[\frac{p}{q}[\omega(E)]^{\frac{q}{p}} + \frac{p}{q(p-1)}[\omega(E)]^{\frac{q}{p}}\right] \\ &= \frac{q}{p}\left[\frac{p}{q} + \frac{p}{q(p-1)}\right][\omega(E)]^{\frac{q}{p}} \\ &= \frac{q}{p}\left[\frac{p(p-1)+p}{q(p-1)}\right][\omega(E)]^{\frac{q}{p}} \\ &= \frac{q}{p}\left[\frac{p^2}{q(p-1)}\right][\omega(E)]^{\frac{q}{p}} \\ &= \left(\frac{p}{p-1}\right)[\omega(E)]^{\frac{q}{p}}.\end{aligned}$$

Finally, we have

$$\|\chi_E\|_{(p,q),\omega} = \left(\frac{p}{p-1}\right)[\omega(E)]^{\frac{q}{p}}.$$

If $1 \leq p < \infty$ and $q = \infty$, then

$$\begin{aligned}
\|\chi_E\|_{(p,\infty),\omega} &= \sup_{t>0} t^{\frac{1}{p}} (\chi_E)^{**}_\omega(t) \\
&= \sup_{t>0} t^{\frac{1}{p}} \min\left(1, \frac{\omega(E)}{t}\right) \\
&= \sup_{\omega(E)>0} \left(\omega(E)\right)^{\frac{1}{p}} \min\left(1, \frac{\omega(E)}{t}\right) \\
&= \sup_{\omega(E)>0} \left(\omega(E)\right)^{\frac{1}{p}} \min(1,1) \\
&= \left[\omega(E)\right]^{\frac{1}{p}}.
\end{aligned}$$

For $p = 1$ and $q < \infty$,

$$\begin{aligned}
\|\chi_E\|_{(\infty,q),\omega} &= \lim_{p\to\infty} \left(\omega(E)\right)^{\frac{1}{p}} \left(\frac{p}{p-1}\right)^{\frac{1}{q}} \\
&= \omega(E) \lim_{p\to\infty} \left(\frac{p}{p-1}\right)^{\frac{1}{q}} \\
&= \omega(E).
\end{aligned}$$

For $p = q = \infty$,

$$\begin{aligned}
\|\chi_E\|_{(\infty,\infty),\omega} &= \sup \min\left(1, \frac{\omega(E)}{t}\right) \\
&= \min\left(1, \omega(E)\right) = 1.
\end{aligned}$$

In summary,

$$\|\chi_E\|_{(p,q),\omega} = \begin{cases}
\left(\frac{p}{p-1}\right)^{1/q} (\omega(E))^{1/p}, & 1 \leq p < \infty, 1 \leq q < \infty, \\
[\omega(E)]^{\frac{1}{p}} & p = 1,\ q < \infty, \\
\infty, & p = 1,\ q < \infty, \\
\omega(E), & p = \infty,\ q < \infty, \\
1, & p = q = \infty.
\end{cases}$$

Proposition 5.10. *Let $(X, \mathcal{A}, \omega d\mu)$ be a σ-finite measure space. Then*

$$\|f\|_{(p,\infty),\omega} = \sup_{t>0} t^{\frac{1}{p}} f^*_\omega(t) = \sup_{t>0} \{\lambda^p D^\omega_f(\lambda)\}^{\frac{1}{p}}.$$

Proof. Let us define

$$C = \sup_{\lambda>0} \{\lambda^p D^\omega_f(\lambda)\}^{\frac{1}{p}}.$$

Then

$$D_f^{\omega}(\lambda) \le \frac{C^p}{\lambda^p}.$$

Choosing $t = \frac{C^p}{\lambda^p}$, we have $\lambda = \frac{C}{t^{\frac{1}{p}}}$ and thus it is clear that

$$f_\omega^*(t) = \inf\{\lambda > 0 : D_f^{\omega}(\lambda) \le t\} \le \frac{C}{t^{\frac{1}{p}}}.$$

Hence

$$t^{\frac{1}{p}} f_\omega^*(t) \le C \quad \text{for all } t > 0,$$

and then

$$\sup t^{\frac{1}{p}} f_\omega^*(t) \le C. \tag{5.10}$$

On the other hand, given $\lambda > 0$, choose $\epsilon > 0$ satisfying $0 < \epsilon < \lambda$, then

$$f_\omega^*(D_f^* - \epsilon) > \lambda,$$

which implies that

$$\begin{aligned} \sup_{t>0} t^{\frac{1}{p}} f_\omega^*(t) &\ge \sup_{\lambda>0} \lambda (D_f^{\omega}(\lambda))^{\frac{1}{p}} \\ &= \sup_{\lambda>0} \{\lambda^p D_f^{\omega}(t)\}^{\frac{1}{p}}. \end{aligned} \tag{5.11}$$

Combining (5.10) and (5.11), we obtain

$$\|f\|_{(p,\infty),\omega} = \sup_{t>0} t^{\frac{1}{p}} f_\omega^*(t) = \sup_{\lambda>0} \{\lambda^p D_f^{\omega}(\lambda)\}^{\frac{1}{p}}.$$

□

Lemma 5.5. *Let $f \in L((p,q), \omega d\mu)$. Then*

$$\sup_{t>0} t^{\frac{1}{p}} f_\omega^{**}(t) \le \|f\|_{(p,q),\omega}.$$

Proof. For $f \in L((p,q), \omega d\mu)$, we can write

$$\begin{aligned} \|f\|_{(p,q),\omega}^q &= \frac{q}{p} \int_0^\infty (t^{\frac{1}{p}} f_\omega^{**}(t))^q \frac{dt}{t} \\ &\ge \frac{q}{p} \int_0^t (f_\omega^{**}(s))^q s^{\frac{q}{p}-1}\, ds \end{aligned}$$

$$\geq \frac{q}{p}[f_\omega^{**}(t)]^q \int_0^t s^{\frac{q}{p}-1}\,ds$$
$$= \frac{q}{p}\frac{p}{q}t^{\frac{q}{p}}[f_\omega^{**}(t)]^q$$
$$= t^{\frac{q}{p}}[f_\omega^{**}(t)]^q.$$

Hence

$$t^{\frac{1}{p}}f_\omega^{**}(t) \leq \|f\|_{(p,q),\omega}.$$

Finally,

$$\sup_{t>0} t^{\frac{1}{p}}f_\omega^{**}(t) \leq \|f\|_{(p,q),\omega}.$$ □

Lemma 5.6.
(I) *If* $|f_n| \uparrow |f|$, *then*

$$\lim_{n\to\infty} D_{f_n}^\omega(\lambda) = D_f^\omega(\lambda).$$

(II) *If* $|f_n| \uparrow |f|$, *then* $(f_n)_\omega^* \uparrow f_\omega^*$
(III) *If* $|f| \leq \liminf |f_n|$, *then* $f_\omega^*(t) \leq \liminf_{n\to\infty}(f_n)_\omega^*$.

Proof. (I) If $|f_n| \uparrow |f|$, then $E_{f_1}(\lambda) \subseteq E_{f_2}(\lambda) \subseteq \cdots$, and hence

$$E_f(\lambda) = \bigcup_{n=1}^\infty E_{f_n}(\lambda).$$

Thus

$$D_f^\omega(\lambda) = \omega(E_f(\lambda)) = \int_{E_f(\lambda)} \omega(x)\,d\mu$$
$$= \int_{\bigcup_{n=1}^\infty E_{f_n}(\lambda)} \omega(x)\,d\mu = \lim_{n\to\infty} \int_{E_{f_n}(\lambda)} \omega(x)\,d\mu$$
$$= \lim_{n\to\infty} \omega(E_{f_n}(\lambda))$$
$$= \lim_{n\to\infty} D_{f_n}^\omega(\lambda).$$

So

$$\lim_{n\to\infty} D_{f_n}^\omega(\lambda) = D_f^\omega(\lambda).$$

(II) By (I), if $|f_n| \uparrow |f|$, then

$$\lim_{n\to\infty} D^{\omega}_{f_n}(\lambda) = D^{\omega}_{f}(\lambda).$$

Let $F_k(t) = D^{\omega}_{f_k}(t)$, then

$$(f_k)^*_\omega = \omega(\{\lambda > 0 : D^{\omega}_{f_n}(\lambda) > t\}) = D^{\omega}_{F_n}(t).$$

Since $D^{\omega}_{f_n}(t) \leq D^{\omega}_{f_{n+1}}(t)$, we have

$$F_n(t) \leq F_{n+1},$$

and thus

$$E_{F_1}(t) \subseteq E_{F_2}(t) \subseteq \cdots,$$

yielding

$$E_{F(t)} = \bigcup_{n=1}^{\infty} E_{F_n}(t).$$

Therefore

$$\lim_{n\to\infty} D^{\omega}_{F_n}(t) = D_F(t).$$

That is

$$\lim_{n\to\infty} (f_n)^*_\omega(t) = f^*_\omega(t).$$

(III) Let $F_n(t) = \inf_{m>n} |f_m(t)|$ and observe that

$$F_n(t) \leq F_{n+1}(t)$$

for all $n \in \mathbb{N}$ and all $t \in X$. Taking $h(t) = \liminf_{n\to\infty} |f_n(t)| = \sup_{n\geq 1} F_n(t)$, we get that $(F_n)^*_\omega \uparrow h_\omega$ as $n \to \infty$, by the fact that $F_n \uparrow h$ and item (I). By hypothesis, we have $|f| \leq h$. Hence

$$f^*_\omega(t) \leq h^*_\omega(t) = \sup_{n\geq 1} (F_n)^*_\omega(t).$$

Since $F_n \leq |f_n|$ for $m \geq n$, it follows that

$$(F_n)^*_\omega(t) \leq \inf_{m\geq n} (f_n)^*_\omega(t).$$

Putting all together, we have

$$\begin{aligned} f_\omega^*(t) \le h_\omega^*(t) &= \sup_{n\ge1}(F_n)_\omega^*(t) \\ &\le \sup_{n\ge1}\inf_{m\ge n}(f_m)_\omega^*(t) \\ &= \liminf_{n\to\infty}(f_n)_\omega^*(t). \end{aligned}$$ □

We now prove the completeness of the $L((p,q),\omega d\mu)$ space.

Theorem 5.7. *The space $L((p,q),\omega d\mu)$ is a Banach space.*

Proof. Let $\{f_n\}_{n\in\mathbb{N}}$ be a Cauchy sequence in $L((p,q),\omega d\mu)$. Then given $\epsilon > 0$, there exists an $n_0 \in \mathbb{N}$ such that

$$\|f_m - f_n\|_{(p,q),\omega} < \epsilon \quad \text{whenever } n_0 \ge m, n. \tag{5.12}$$

Applying Lemma 5.5, Proposition 5.10, and Proposition 5.9(c), we have

$$\begin{aligned} \sup_{\lambda>0}\{\lambda^p D^\omega_{f_m-f_n}(\lambda)\}^{\frac{1}{p}} &= \sup_{t>0} t^{\frac{1}{p}}(f_m - f_n)_\omega^*(t) \\ &\le \sup_{t>0} t^{\frac{1}{p}}(f_m - f_n)_\omega^{**}(t) \\ &\le \|f_m - f_n\|_{(p,q),\omega} < \epsilon \end{aligned}$$

whenever $n_0 \ge m, n$.

This means that $\{f_n\}$ is a Cauchy sequence in measure with respect to $\omega d\mu$. Therefore, we can now apply F. Riesz's theorem and conclude that there exists an $\mathcal{A}$-measurable function f such that $\{f_n\}$ converges in measure (with respect $\omega d\mu$) to f. This implies again by a theorem of F. Riesz that there is a subsequence $\{f_{n_k}\}_{k\in\mathbb{N}}$ of $\{f_n\}_{n\in\mathbb{N}}$ which converges to f.

By (5.12), $f_{n_k} - f_{n_0}$ (if $n > n_0$) converge to $f - f_{n_0}$ $\omega d\mu$ a. e. on X. It follows now by Lemma 5.5(II) that

$$(f - f_{n_0})_\omega^*(t) \le \liminf_{n\to\infty}(f_{n_k} - f_{n_0})_\omega^*(t)$$

for all $t > 0$. Using Fatou's lemma, we have

$$(f - f_{n_0})_\omega^{**}(t) \le \liminf_{n\to\infty}(f_{n_k} - f_{n_0})_\omega^{**}(t)$$

for all $t > 0$. One more time, by Fatou's lemma, we have

$$\|f - f_{n_0}\|_{(p,q),\omega} = \left(\frac{q}{p}\int_0^\infty (t^{\frac{1}{p}}(f - f_{n_0})_\omega^{**}(t))^q \frac{dt}{t}\right)^{\frac{1}{q}}$$
$$\leq \left(\int_0^\infty \left(t^{\frac{1}{p}} \liminf_{k\to\infty}(f_{n_k} - f_{n_0})_\omega^{**}(t)\right)^q \frac{dt}{t}\right)^{\frac{1}{q}}$$
$$\leq \liminf_{k\to\infty}\left(\int_0^\infty (t^{\frac{1}{p}}(f_{n_k} - f_{n_0})_\omega^{**}(t))^q \frac{dt}{t}\right)^{\frac{1}{q}}$$
$$\leq \liminf_{k\to\infty} \|f_{n_k} - f_{n_0}\|_{(p,q),\omega} < \epsilon$$

whenever $n_k > n_0$. Since

$$f = (f - f_{n_0}) + f_{n_0} \in L((p,q), \omega d\mu),$$

this proves that $L((p,q), \omega d\mu)$ is complete. □

The following result, due to G. H. Hardy, will be useful in proving that the norms $|\cdot\|_{p,q,\omega}$ and $\|\cdot\|_{(p,q),\omega}$ are equivalent.

Theorem 5.8 (G. H. Hardy). *If f is a nonnegative valued-measurable function on $[0,\infty)$ and if q and r are two numbers satisfying $1 \leq q < \infty$ and $0 < r < \infty$, then*

$$\int_0^\infty \left(\int_0^t f(s)\,ds\right)^q t^{-r-1}\,dt \leq \left(\frac{q}{r}\right)^q \int_0^\infty (sf(s))^q s^{-r-1}\,ds.$$

Proof. If $q = 1$, by Fubini's theorem, we have

$$\int_0^\infty \left(\int_0^t f(s)\,ds\right) t^{-r-1}\,dt = \int_0^\infty \int_s^\infty t^{-r-1} f(s)\,dtds$$
$$= \frac{1}{r}\int_0^\infty [sf(s)]s^{-r-1}\,ds$$
$$= \left(\frac{q}{r}\right)^q \int_0^\infty (sf(s))^q s^{-r-1}\,ds.$$

Now, suppose that $q > 1$ and let p be the conjugate exponent of q. Then by Hölder's inequality with respect to the measure $s^{\frac{r}{q}-1}\,ds$, we have

$$\begin{aligned}\left(\int_0^t f(s)\,ds\right) &= \left(\int_0^t f(s)s^{1-\frac{r}{q}}s^{\frac{r}{q}-1}\,ds\right)^q\\ &\le \left(\int_0^t [f(s)]^q s^{q-r}s^{\frac{r}{q}-1}\,ds\right)^q\left(\int_0^t s^{\frac{t}{q}-1}\,ds\right)^{\frac{q}{p}}\\ &= \left(\frac{q}{r}t^{\frac{r}{q}}\right)^{\frac{q}{p}}\int_0^t [f(s)]^q s^{q-r}s^{\frac{r}{q}-1}\,ds\\ &= \left(\frac{q}{r}\right)^{\frac{q}{p}}t^{\frac{r}{p}}\int_0^t [f(s)]^q s^{q-r}s^{\frac{r}{q}-1}\,ds.\end{aligned}$$

Integrating both sides from zero to infinity and using Fubini's theorem, we have

$$\begin{aligned}\int_0^\infty\left(\int_0^t f(s)\,ds\right)^q t^{-r-1}\,dt &\le \left(\frac{q}{r}\right)^{\frac{q}{p}}\int_0^\infty t^{-r-1+r(1-\frac{1}{q})}\int_0^t [f(s)]^q s^{q-r}s^{\frac{r}{q}-1}\,dsdt\\ &= \left(\frac{q}{r}\right)^{\frac{q}{p}}\int_0^\infty [f(s)]^q s^{q-r}s^{\frac{r}{q}-1}\left(\int_s^\infty t^{-1-\frac{r}{q}}\right)ds\\ &= \left(\frac{q}{r}\right)^{\frac{q}{p}+1}\int_0^\infty [sf(s)]^q s^{-r+\frac{r}{q}-1}s^{-\frac{r}{q}}\,ds\\ &= \left(\frac{q}{r}\right)^q\int_0^\infty [sf(s)]^q s^{-r-1}\,ds.\end{aligned}$$

Hence

$$\int_0^\infty\left(\int_0^t f(s)\,ds\right)^q t^{-r-1} \le \left(\frac{q}{r}\right)^q\int_0^\infty (sf(s))^q s^{-r-1}\,ds. \qquad \square$$

We are ready to prove the equivalence between the $|\cdot\|_{p,q,\omega}$ and $\|\cdot\|_{(p,q),\omega}$ norms.

Theorem 5.9. *If* $1 < p \le \infty$ *and* $1 \le q \le \infty$*, then*

$$\|f\|_{p,q,\omega} \le \|f\|_{(p,q),\omega} \le \frac{p}{p-1}\|f\|_{p,q,\omega}.$$

Proof. Note that, by Hardy's inequality,

$$\begin{aligned}
\|f\|^q_{(p,q),\omega} &= \frac{q}{p}\int_0^\infty [t^{\frac{1}{p}} f^{**}_\omega(t)]^q \frac{dt}{t} \\
&= \frac{q}{p}\int_0^\infty t^{\frac{q}{p}-1}\left(\frac{1}{t}\int_0^t f^*_\omega(s)\,ds\right)^q dt \\
&= \frac{q}{p}\int_0^\infty t^{\frac{q}{p}-1}\left(\int_0^t f^*_\omega(s)\,ds\right)^q t^{\frac{q}{p}-q-1}\,dt \\
&= \frac{q}{p}\int_0^\infty \left(\int_0^t f^*_\omega(s)\,ds\right)^q t^{-(q-\frac{q}{p})-1}\,dt \\
&\le \left(\frac{q}{q-\frac{q}{p}}\right)^q \frac{q}{p}\int_0^\infty (sf^*_\omega(s))^q s^{-(q-\frac{q}{p})-1}\,ds \\
&= \left(\frac{q}{\frac{q(p-1)}{p}}\right)^q \frac{q}{p}\int_0^\infty s^q s^{-q}(s^{\frac{1}{p}} f^*_\omega(s))^q \frac{ds}{s} \\
&= \left(\frac{p}{p-1}\right)^q \frac{q}{p}\int_0^\infty (s^{\frac{1}{p}} f^*_\omega(s))^q \frac{ds}{s} \\
&= \left(\frac{p}{p-1}\right)^q \|f\|^q_{p,q,\omega}.
\end{aligned}$$

Hence

$$\|f\|_{(p,q),\omega} \le \frac{p}{p-1}\|f\|_{p,q,\omega}. \tag{5.13}$$

Since $f^*_\omega(t) \le f^{**}_\omega(t)$, we have

$$\|f\|_{p,q,\omega} \le \|f\|_{(p,q),\omega}. \tag{5.14}$$

Using (5.13) and (5.14), we arrive at

$$\|f\|_{p,q,\omega} \le \|f\|_{(p,q),\omega} \le \frac{p}{p-1}\|f\|_{p,q,\omega}. \qquad \square$$

5.5 *K*-method

A pair (X_0, X_1) of Banach spaces X_0 and X_1 is called a compatible couple if there is some Hausdorff topological vector space, say $\mathcal{H}$, in which each of X_0 and X_1 is continuously embedded. Further, the sum $X_0 + X_1$ is a Banach space with the norm

$$\|f\|_{X_0+X_1} := \inf\{\|f_0\|_{X_0} + \|f_1\|_{X_1} : f = f_0 + f_1, f_0 \in X_0, f_1 \in X_1\}.$$

The pair (L_1, L_∞) is a compatible couple because both L_1 and L_∞ are continuously embedded in the Hausdorff space $(\mathcal{M}(0,\infty), \mathcal{L})$ of measurable functions that are finite a. e.

Definition 5.6. Let (X_0, X_1) be a compatible couple of Banach spaces. The K-functional is defined for each $f \in X_0 + X_1$ and $t > 0$ by

$$K(f, t, X_0, X_1) := \inf\{\|f_0\|_{X_0} + t\|f_1\|_{X_1} : f = f_0 + f_1\},$$

where the infimum is over all representations $f = f_0 + f_1$ of f with $f_0 \in X_0$ and $f_1 \in X_1$.

For more details on the K-method, see [2]. The Peetre K functional is constructed from this expression by introducing a positive weighting factor t as follows.

Proposition 5.11. *Let $f \in X_0 + X_1$. Then $K(f, t, X_0, X_1)$ satisfies the following properties:*
(I) $K(f, s, X_0, X_1) \le K(f, t, X_0, X_1)$, *for* $s \le t$;
(II) $K(f, \cdot, X_0, X_1)$ *is a nonnegative concave function of* $t > 0$;
(III) $K(f, t, X_0, X_1) = tK(f, \frac{1}{t}, X_0, X_1)$.

Proof. (I) Let $s \le t$. Then

$$\|f_0\|_{X_0} + s\|f_1\|_{X_1} \le \|f_0\|_{X_0} + t\|f_1\|_{X_1}$$

for $f = f_0 + f_1, f_0 \in X_0, f_1 \in X_1$. Thus

$$K(f, s, X_0, X_1) \le K(f, t, X_0, X_1).$$

(II) Let $t = (1-\lambda)t_1 + \lambda t_2$ with $0 < \lambda < 1$, $t_1, t_2 > 0$. Then

$$\begin{aligned}
&(1-\lambda)K(f, t_1, X_0, X_1) + \lambda K(f, t_2, X_0, X_1)\\
&\le (1-\lambda)(\|f_0\|_{X_0} + t_1\|f_1\|) + \lambda(\|f_0\|_{X_0} + t_2\|f_1\|_{X_1})\\
&= \|f_0\|_{X_0} + ((1-\lambda)t_1 + \lambda t_2)\|f_1\|_{X_1}\\
&= \|f_0\|_{X_0} + t\|f_1\|_{X_1}.
\end{aligned}$$

Taking the infimum over all such decompositions $f = f_0 + f_1$ of f, we therefore obtain

$$(1-\lambda)K(f, t_1, X_0, X_1) + \lambda K(f, t_2, X_0, X_1) \le K(f, t, X_0, X_1).$$

Hence $t \to K(f, t, X_0, X_1)$ is concave, as desired.

(III)

$$\begin{aligned}K(f,t,X_0,X_1) &= \inf\{\|f_0\|_{X_0} + t\|f_1\|_{X_1} : f = f_0 + f_1 \in X_0, f_1 \in X_1\}\\ &= t\inf\left\{\frac{1}{t}\|f_0\|_{X_0} + \|f_1\|_{X_1} : f = f_0 + f_1 \in X_0, f_1 \in X_1\right\}\\ &= t\inf\left\{\|f_1\|_{X_1} + \frac{1}{t}\|f_0\|_{X_0} : f = f_0 + f_1 \in X_0, f_1 \in X_1\right\}\\ &= tK\left(f, \frac{1}{t}, X_0, X_1\right).\end{aligned}$$

Thus $K(f,t,X_0,X_1) = tK(f, \frac{1}{t}, X_1, X_0)$. □

Theorem 5.10. *If $0 < p < \infty$ and f is a measurable function on X, then*

$$K(f,t,\Lambda_X^p(\omega), L_\infty(X)) \sim K(f^*,t,L_p(\omega),L_\infty(X)), \quad t > 0.$$

Proof. If $f = f_0 + f_1$ with $f_0 \in \Lambda_X^p(\omega)$ and $f_1 \in L_\infty(X)$, then we have that

$$\begin{aligned}f^*(s) &\le f^*(s) + f_1^*(0)\\ &= f_0^*(s) + \|f_1\|_{L_\infty(X)}, \quad s > 0.\end{aligned}$$

Hence

$$\begin{aligned}K(f^*,t,L_p(\omega),L_\infty) &\le \|f_0^*\|_{L_p(\omega)} + t\|f_1\|_{L_\infty(X)}\\ &= \|f_0\|_{\Lambda_X^p(\omega)} + t\|f_1\|_{L_{\infty(X)}}.\end{aligned}$$

Taking the infimum over all decompositions

$$f = f_0 + f_1 \in \Lambda_X^p(\omega) + L_\infty(X),$$

we obtain

$$K(f^*,t,L_p(\omega),L_\infty(\omega)) \le K(f,t,\Lambda_X^p(\omega),L_\infty(X)).$$

To prove the converse inequality, if f is measurable and $t > 0$, let

$$f_0 = \left(f - (f^*)_\omega^*(t^p)\frac{f}{|f|}\right)\chi_{\{|f|>(f^*)_\omega^*(t^p)\}}$$

and

$$f_1 = f - f_0.$$

Then

$$(f_0^*)_\omega^* = ((f^*)_\omega^* - (f^*)_\omega^*(t^p))\chi_{[0,t^p]},$$

while

$$f_1^* \le (f^*)_\omega^*(t^p).$$

Since $f = f_0 + f_1$, we have that

$$\begin{aligned} K(f,t,\Lambda_X^p, L_\infty(X)) &\le \|f_0\|_{\Lambda_X^p(\omega)} + t\|f_1\|_{L_\infty(X)} \\ &\le \|f_0^*\|_{L_p(\omega)} + t(f^*)_\omega(t^p) \\ &= \|(f_0^*)_\omega^*\|_{L_p} + t(f^*)_\omega(t^p) \\ &= \left(\int_0^{t^p} ((f^*)_\omega^*(s) - (f^*)_\omega^*(t^p))^p\, ds\right)^{\frac{1}{p}} + \left(\int_0^{t^p} [(f^*)_\omega^*(t^p)]^p\, ds\right)^{\frac{1}{p}}. \end{aligned}$$

Now, we shall consider two cases:
(I) $0 < p \le 1$;
(II) $p > 1$.

Case (I). We have

$$\begin{aligned} &\left(\int_0^{t^p} ((f^*)_\omega^*(s) - (f^*)_\omega^*(t^p))^p\, ds\right)^{\frac{1}{p}} + \left(\int_0^{t^p} [(f^*)_\omega^*(t^p)]^p\, ds\right)^{\frac{1}{p}} \\ &\quad \le \left(\int_0^{t^p} ((f^*)_\omega^*(s) - (f^*)_\omega^*(t^p) + (f^*)_\omega^*(t^p))^p\, ds\right)^{\frac{1}{p}} \\ &\quad = \left(\int_0^{t^p} ((f^*)_\omega^*(s))^p\, ds\right)^{\frac{1}{p}}. \end{aligned}$$

Case (II). Since $(f^*)_\omega^*(t^p) \le (f^*)_\omega^*(s)$, if $0 < s < t^p$, we get

$$\left(\int_0^{t} ((f^*)_\omega^*(s) - (f^*)_\omega^*(t^p))^p\, ds\right)^{\frac{1}{p}} + \left(\int_0^{t} [(f^*)_\omega^*(t^p)]^p\, ds\right)^{\frac{1}{p}} \le 2\left(\int_0^{t} [(f^*)_\omega^*(t^p)]^p\, ds\right)^{\frac{1}{p}}.$$

In both cases, the latter expression is equivalent to $K(f^*, t, L_p(\omega), L_\infty(X))$. □

6 BMO spaces

The space of functions of bounded mean oscillation, or BMO, naturally arises as the class of functions whose deviation from their means over cubes is bounded. Note that L_∞ functions have this property, but there exist unbounded functions with bounded mean oscillation, for instance, the function $\log|x|$ belongs to BMO, but it is not bounded. The BMO space shares similar properties with the L_∞ space and it often serves as a substitute for it. The space of functions with bounded mean oscillation BMO is well known for its several applications in real analysis, harmonic analysis, and partial differential equations.

It is worth to say that this chapter is based on [18].

Definition 6.1. The BMO space is the set of all locally integrable functions f such that

$$\|f\|_{\mathrm{BMO}} := \sup_Q \int_Q \left|f(x) - f_Q\right| dx < \infty, \tag{6.1}$$

where

$$f_Q = \frac{1}{m(Q)} \int_Q f(y)\, dy,$$

$m(Q)$ is the Lebesgue measure of Q, and Q is a cube in $\mathbb{R}^n$, with sides parallel to the coordinate axes.

In [38] Garnet and Jones gave upper and lower bounds for the distance

$$\operatorname{dist}(f, L_\infty) = \inf_{g \in L_\infty} \|f - g\|_{\mathrm{BMO}}. \tag{6.2}$$

The bounds were expressed in terms of one constant in John–Nirenberg inequality. John and Nirenberg proved in [49] that $f \in \mathrm{BMO}$ if and only if there exist $\epsilon > 0$ and $\lambda_0 = \lambda_0(\epsilon)$ such that

$$\sup_Q \frac{1}{m(Q)} \{x \in Q : \left|f(x) - f_Q\right| > \lambda\} \le e^{-\frac{\lambda}{\epsilon}}, \tag{6.3}$$

whenever $\lambda > \lambda_0 = \lambda_0(f, \epsilon)$. Indeed, when $f \in \mathrm{BMO}$, (6.3) holds with $\epsilon = C\|f\|_{\mathrm{BMO}}$, where the constant C depends only on the dimension.

Specifically, setting

$$\epsilon(f) = \inf\{\epsilon > 0 : f \text{ satisfies (6.3)}\},$$

Garnett and Jones proved that

$$A_1 \epsilon(f) \le \operatorname{dist}(f, L_\infty) \le A_2 \epsilon(f),$$

https://doi.org/10.1515/9783112223246-006

where A_1 and A_2 are constants depending only on the dimension. Also, they observed that $\operatorname{dist}(f, L_\infty)$ can be related to the growth of

$$\sup_Q \left(\frac{1}{m(Q)} \int_Q |f(x) - f_Q|^p \, dx \right)^{\frac{1}{p}}$$

as $p \to \infty$. This is because

$$\frac{\epsilon(f)}{\epsilon} = \lim_{p \to \infty} \frac{1}{p} \left(\sup_Q \frac{1}{m(Q)} \int_Q |f(x) - f_Q|^p \, dx \right)^{\frac{1}{p}}. \tag{6.4}$$

Our next goal is to extend (6.4) to BMO_φ^p (see Section 6.1 and Theorem 6.1 on spaces of homogeneous type).

6.1 Spaces of homogeneous type and $\mathrm{BMO}_\varphi^{(p)}(X)$ spaces

Let us recall the notion of a *space of homogeneous type.*

Definition 6.2. A quasimetric d on a set X is a function $d : X \times X \to [0, \infty)$ with the following properties:
1. $d(x,y) = 0$ if and only if $x = y$;
2. $d(x,y) = d(y,x)$ for all $x, y \in X$;
3. There exists a constant K such that

$$d(x,y) \le K\left[d(x,z) + d(z,y)\right],$$

for all $x, y, z \in X$.

A quasimetric defines a topology in which the balls

$$B(x,r) = \{y \in X : d(x,y) < r\}$$

form a base. These balls may not be open sets. However, from a given quasimetric d, it is easy to build an equivalent quasimetric d' such that the d'-quasimetric balls are open (the existence of d' has been proved by using topological arguments in [56]). So, without loss of generality, we assume that the quasimetric balls are open. A general method of constructing families $\{B(x, \delta)\}$ is in terms of a quasimetric.

Definition 6.3. A space of homogeneous type (X, d, μ) is a set X with a quasimetric d and a Borel measure μ, finite on bounded sets and such that, for some absolute positive constant A, the following doubling property holds:

$$\mu\big(B(x, 2r)\big) \le A\mu\big(B(x, r)\big)$$

for all $x \in X$ and $r > 0$.

Let us take a look at some examples of spaces of homogeneous type.

Example 6.1. Let $X \subset \mathbb{R}^n$, $X = \{0\} \cup \{x : |x| = 1\}$, put in X the Euclidean distance and the following measure μ: μ is the usual surface measure on $\{x : |x| = 1\}$ and $\mu(\{0\}) = 1$. Then μ is doubling, so (X, d, μ) is a homogeneous space.

Example 6.2. In $\mathbb{R}^n$, let C_k $(k = 1, 2, \ldots)$ be the point $(k^k + \frac{1}{2}, 0, \ldots, 0)$, for $k \geq 2$, let B_k be the ball $B(C_k, \frac{1}{2})$, and $B_1 = B(0, \frac{1}{2})$. Let $X = \bigcup_{k=1}^{\infty} B_k$ with the Euclidean distance and the measure μ such that $\mu(B_k) = 2^k$ and, on each ball B_k, μ being uniformly distributed.

Claim 1. μ satisfies the doubling condition. Let $B_r = B(P, r)$ with $P = (P_1, \ldots, P_n)$ and $r > 0$.

Case 1. Assume for some k, $B_k \subset B_r$ and let $k_0 = \max\{k : B_k \subset B_r\}$. Then certainly $P_1 + r \leq b_{k_0+1} = (k_0 + 1)^{k_0+1} + 1$ and $\mu(B_r) \geq 2^{k_0}$. But, then

$$\begin{aligned} P_1 + 2r &\leq 2\big((k_0 + 1)^{k_0+1} + 1\big) \\ &\leq (k_0 + 2)^{k_0+2} = a_{k_0+2}. \end{aligned}$$

Therefore $B_{2r} \subset B_{a_{k_0+2}}(0) \equiv B_0$. But

$$\mu(B_0) = \sum_{K=0}^{k_0+1} 2^k \leq 2^{k_0+2} \leq 4\mu(B_r).$$

Hence the doubling condition holds with $A = 4$.

Case 2. If for all k, $B_k \not\subset B_r$, then $r \leq 1$, so that B_r and B_{2r} intersect only one ball B_k. Then the doubling condition holds.

Now, we are ready to define the *bounded* (φ, p) *mean oscillation* spaces.

Definition 6.4. Let X be a space of homogeneous type, φ a nonnegative function on $[0, \infty)$, and $1 \leq p < \infty$. The space of functions of bounded (φ, p) mean oscillation, $\mathrm{BMO}_\varphi^{(p)}(X)$, is the set of all locally μ-integrable functions $f : X \to \mathbb{R}$ such that

$$\sup\left(\frac{1}{\mu(B)[\varphi(\mu(B))]^p} \int_B |f(x) - f_B|^p \, d\mu(x)\right)^{\frac{1}{p}} < \infty,$$

where the supremum is taken over all balls $B \subset X$, and

$$f_B = \frac{1}{\mu(B)} \int_B f(y) \, d\mu.$$

Remark 6.1. It is not hard to check that the expression

$$\|f\|_{\mathrm{BMO}_\varphi^p} = \sup_B \left(\frac{1}{\mu(B)[\varphi(\mu(B))]^p} \int_B |f(x) - f_B|^p \, d\mu(x)\right)^{\frac{1}{p}} \tag{6.5}$$

defines a norm on $\mathrm{BMO}_\varphi^{(p)}(X)$. For $\varphi \equiv 1$ and $p = 1$, the $\|\cdot\|_{\mathrm{BMO}_\varphi^p}$ norm coincides with the $\|\cdot\|_{\mathrm{BMO}}$ norm in (6.1).

A warning about notation. If there is no risk of confusion, we write $\mathrm{BMO}_\varphi^{(p)}$ for $\mathrm{BMO}_\varphi^{(p)}(X)$. Also, when $p = 1$, we denote $\mathrm{BMO}_\varphi^{(1)}(X)$ as $\mathrm{BMO}_\varphi(X)$ (or just BMO_φ) and write $\|f\|_{\mathrm{BMO}_\varphi}$ instead of $\|f\|_{\mathrm{BMO}_\varphi^1}$.

6.2 John–Nirenberg inequality on a homogeneous type space

We establish and demonstrate the John–Nirenberg inequality, employing reasoning analogous to that presented in [57].

Theorem 6.1. *There exist two positive constants β and b such that for any $f \in \mathrm{BMO}_\varphi(X)$ and any ball $B \subset X$, one has*

$$\mu(\{x \in S : |f - f_B| > \lambda\}) \leq \beta \exp\{-b\lambda/\|f\|_{\mathrm{BMO}_\varphi}\}\mu(B). \tag{6.6}$$

Proof. We follow the standard stopping time argument, that is, we assume that λ is large enough and fix some λ_1. Then we study the sets $\{x \in S : |f(x) - f_S| \leq \lambda_1\}$, $\{x \in S : |f(x) - f_S| \leq 2\lambda_1\}$ up to

$$\{x \in S : |f(x) - f_S| \leq m\lambda_1 \sim \lambda\}.$$

To show (6.6), we assume $\|f\|_\varphi = 1$ and fix $S = B(a, R)$. We define a maximal operator associated to S (if we replace S by another ball, then the maximal operator changes)

$$M_S f(x) = \sup_{\substack{B \text{ Ball}, x \in B, \\ B \subset B(a,R)}} \left\{ \frac{1}{\varphi(\mu(B))\mu(B)} \int_B |f(y) - f_S| \, d\mu(y) \right\}.$$

Using a Vitali-type covering lemma, one can prove that

$$\mu(\{x : M_S f(x) > t\}) \leq \frac{A}{t}\mu(S),$$

where A is a constant that depends only on K and k_2 but not on S. Take $\lambda_0 > A$ and consider the open set $U = \{x : M_S f(x) > \lambda_0\}$. We have

$$\mu(U \cap S) \leq \frac{A}{\lambda_0}\mu(S) < \mu(S),$$

and therefore $S \cap U^c \neq 0$. Define

$$r(x) = \frac{1}{5K} \operatorname{dist}(x, U^c).$$

If $x, y \in S$, then $d(x, y) \leq 2KR$. Since $S \cap U^c \neq \emptyset$, if $x \in S$, we have $r(x) \leq \frac{2KR}{(5K)} = \frac{2R}{5}$.

Clearly,

$$U \cap S \subset \bigcup_{x \in U \cap S} B(x, r(x)) \subset U.$$

Again by a Vitali-type covering lemma, we can select a finite or countable sequence of disjoint balls $\{B(x_j, r_j)\}$ such that $r_j = r_j(x)$ and

$$U \cap S \subset \bigcup_j B(x_j, 4Kr_j(x)) \subset U.$$

On the other hand, $B(x_j, 6Kr_j) \cap U^c \neq \emptyset$ and $B(x_j, 6Kr_j) \subset B(a, aR)$ because $6Kr_j \leq \frac{12KR}{5}$. Thus, we get

$$\frac{1}{\varphi(\mu(B(x_j, 6Kr_j)))\mu(B(x_j, 6Kr_j))} \int_{B(x_j, 6Kr_j)} |f - f_S| \, d\mu \leq \lambda_0$$

and, consequently, writing $S_j = B(x_j, 4Kr_j)$, we obtain

$$\begin{aligned}
|f_S - f_{S_j}| &\leq \frac{1}{\mu(S_j)} \int_{S_j} |f - f_S| \, d\mu \\
&\leq \frac{\varphi(S_j)k_2^2}{\mu(B(x_j, Kr_j))}\lambda_0 := \lambda_1
\end{aligned}$$

because μ is a doubling measure.

By differentiation theorem, $|f(x) - f_S| \leq \lambda_0$ for μ-a. e. $x \in S \setminus \bigcup_j S_j$. Moreover,

$$\begin{aligned}
\sum \mu(S_j) &\leq k_2 \sum_j \mu(B(x_j, 2Kr_j)) \\
&\leq C \sum_j \mu(B(x_j, r_j)) \\
&\leq C\mu(U) \\
&\leq \frac{CA}{\lambda_0}\mu(S).
\end{aligned}$$

Now, we do the same construction for each S_j. Again $|f(x) - f_S| \leq \lambda_0$ for μ-a. e. $x \in S_j \setminus \bigcup_i S_i^{(2)}$ and therefore for these points

$$\begin{aligned}
|f(x) - f_S| &\leq |f(x) - f_{S_j}| + |f_{S_j} - f_S| \\
&\leq \lambda_0 + \frac{\varphi(S_j)k_2^2}{\mu(B(x_j, Kr_j))}\lambda_0 \\
&\leq \frac{2\varphi(S_j)k_2^2}{\mu(B(x_j, Kr_j))}\lambda_0.
\end{aligned}$$

Taking $\lambda_0 = 2CA$, it is clear that

$$\mu\Big(\bigcup_k S_k^{(2)}\Big) \leq \sum_j \frac{CA}{\lambda_0}\mu(S_j)$$
$$\leq \Big(\frac{CA}{\lambda_0}\Big)^2 \mu(S) = 2^{-2}\mu(S).$$

Continuing in this manner, we get for $N = 1, 2, \dots$ a family of balls $\{S_j^N\}$ such that

$$\mu\Big(\bigcup_j S_j^N\Big) \leq 2^{-N}\mu(S).$$

Finally,

$$\mu(\{x \in S : |f(x) - f_S| > \lambda\}) \leq \mu(\{x \in S : |f(x) - f_S| > N\lambda_1\})$$
$$\leq \mu\Big(\bigcup_j S_j^N\Big) \leq 2^{-N}\mu(S) = e^{-b\lambda}\mu(S).$$

This completes the proof. □

6.3 Completeness of BMO_φ spaces

In this section we state some simple lemmas. The first is showed by elementary calculations.

Lemma 6.1. *Let B_0 and B_1 be two balls such that $B_0 \subset B_1$ and $f \in \mathrm{BMO}_\varphi$. Then there exists a constant C depending on B_0 and B_1 such that*

$$|f_{B_0} - f_{B_1}| \leq C\|f\|_{\mathrm{BMO}_\varphi}.$$

Proof. Indeed,

$$|f_{B_0} - f_{B_1}| = \Big|\frac{1}{\mu(B_0)}\int_{B_0}(f(y) - f_{B_1})\,d\mu(y)\Big|$$
$$\leq \frac{1}{\mu(B_0)}\int_{B_0}|f(y) - f_{B_1}|\,d\mu(y)$$
$$= \frac{\mu(B_1)}{\mu(B_0)}\,\frac{\varphi(\mu(B))}{\varphi(\mu(B))\mu(B_1)}\int_{B_1}|f(y) - f_{B_1}|\,d\mu(y)$$
$$\leq \frac{\mu(B_1)\varphi(\mu(B))}{\mu(B_0)}\|f\|_{\mathrm{BMO}_\varphi}.$$

This completes the proof. □

The result below tells us about the equivalence of the $\|\cdot\|_{\mathrm{BMO}_\varphi}$ and $\|\cdot\|_{\mathrm{BMO}^p_\varphi}$ norms.

Lemma 6.2 (John–Nirenberg type lemma). *Let $f \in \mathrm{BMO}^{(p)}_\varphi(X)$, $1 \le p < \infty$, then there exists a constant C_p such that*

$$\|f\|_{\mathrm{BMO}_\varphi} \le \|f\|_{\mathrm{BMO}^p_\varphi} \le C_p\|f\|_{\mathrm{BMO}_\varphi}.$$

Proof. By Hölder's inequality, we have

$$\frac{1}{\varphi(\mu(B))\mu(B)} \int_B |f(y) - f_B|\, d\mu(y) \le \sup_B \left(\frac{1}{[\varphi(\mu(B))]^p \mu(B)} \int_B |f(y) - f_B|^p\, d\mu(y) \right)^{\frac{1}{p}}$$

for any ball, thus

$$\|f\|_{\mathrm{BMO}_\varphi} \le \|f\|_{\mathrm{BMO}^{(p)}_\varphi}.$$

On the other hand,

$$\int_B |f(y) - f_B|^p\, d\mu(y) \le \int_0^\infty p\lambda^{p-1}\mu(\{x \in B : |f(x) - f_B| > \lambda\})\, d\lambda.$$

By Theorem 6.1, we obtain

$$\int_B |f(y) - f_B|^p\, d\mu(y) \le \int_0^\infty p\lambda^{p-1} \exp(-b\lambda/\|f\|_{\mathrm{BMO}_\varphi})\mu(B)\, d\lambda.$$

Therefore,

$$\frac{1}{[\varphi(B)]^p\mu(B)} \int_B |f(y) - f_B|^p\, d\mu(y) \le p\Gamma(p)C\|f\|_{\mathrm{BMO}_\varphi},$$

and thus

$$\|f\|_{\mathrm{BMO}^{(p)}_\varphi} \le C_p\|f\|_{\mathrm{BMO}_\varphi},$$

finishing the proof. □

We now prove the completeness of the BMO^p_φ spaces.

Theorem 6.2. BMO^p_φ *equipped with the norm* (6.4) *is a Banach space.*

Proof. Let us take B_1 to be the unit ball centered at the origin. Let $f_k \in \mathrm{BMO}_\varphi^{(p)}$, for each $k = 1, 2, 3, \ldots$, such that

$$\sum_{k=1}^{\infty} \|f_k\|_{\mathrm{BMO}_\varphi^{(p)}} < \infty,$$

and assume that

$$\int_{B_1} f_k(y)\, d\mu(y) = 0. \tag{6.7}$$

Let B be any ball in X and let B_2 be a ball that contains both B_1 and B. Then

$$\sum_{k=1}^{\infty} \left(\frac{1}{\mu(B)} \int_B |f_k(y)|^p \, d\mu(y) \right)^{\frac{1}{p}} = \left(\frac{\mu(B_2)}{\mu(B)} \right)^{\frac{1}{p}} \sum_{k=1}^{\infty} \left(\frac{1}{\mu(B_2)} \int_{B_2} |f_k(y)|^p \, d\mu(y) \right)^{\frac{1}{p}}.$$

By Minkowski inequality and (6.7), we have

$$\begin{aligned}
&\sum_{k=1}^{\infty} \left(\frac{1}{\mu(B)} \int_B |f_k(y)|^p \, d\mu(y) \right)^{\frac{1}{p}} \left(\frac{[\varphi(\mu(B_2))]^p \mu(B_2)}{\mu(B)} \right)^{\frac{1}{p}} \\
&\leq \sum_{k=1}^{\infty} \left(\frac{1}{[\varphi(\mu(B_2))]^p \mu(B_2)} \int_{B_2} |f_k(y) - f_{B_2}|^p \, d\mu(y) \right)^{\frac{1}{p}} \\
&\quad + \left(\frac{\mu(B_2)}{\mu(B)} \right)^{\frac{1}{p}} \sum_{k=1}^{\infty} \left(\frac{1}{\mu(B_2)} \int_{B_2} |(f_k)_{B_2} - (f_k)_{B_1}|^p \, d\mu(y) \right)^{\frac{1}{p}} \\
&\leq \left(\frac{[\varphi(\mu(B_2))]^p \mu(B_2)}{\mu(B)} \right)^{\frac{1}{p}} \sum_{k=1}^{\infty} \left[\|f_k\|_{\mathrm{BMO}_\varphi^{(p)}} + \left(\frac{\mu(B_2)}{\mu(B)} \right)^{\frac{1}{p}} |(f_k)_{B_2} - (f_k)_{B_1}| \right].
\end{aligned}$$

By Lemma 6.2, is easy to see that

$$\sum_{k=1}^{\infty} \left(\frac{1}{\mu(B)} \int_B |f_k(y)|^p \, d\mu(y) \right)^{\frac{1}{p}} \leq \left(\frac{\mu(B_2)}{\mu(B)} \right)^{\frac{1}{p}} \sum_{k=1}^{\infty} (1 + [\varphi(\mu(B_2))]^p) \|f_k\|_{\mathrm{BMO}_\varphi^p}.$$

Therefore,

$$\sum_{k=1}^{\infty} \left(\frac{1}{\mu(B)} \int_B |f_k(y)|^p \, d\mu(y) \right)^{\frac{1}{p}} < \infty.$$

This means

$$\left(\frac{1}{\mu(B)}\right)^{\frac{1}{p}} \sum_{k=1}^{\infty} \|f_k\|_{L_p} < \infty, \tag{6.8}$$

and from (6.8) we obtain

$$f = \lim_{m\to\infty} \sum_{k=1}^{m} f_k \quad \text{a. e.}$$

For $f \in L_p$, clearly, $f_B = \sum_{k=1}^{\infty} (f_k)_B$.

Finally, we want to show that:

(a) $f \in \mathrm{BMO}_\varphi^{(p)}(X)$,

(b) $\| \sum_{k=1}^{m} f_k - f\|_{\mathrm{BMO}_\varphi^{(p)}} \to 0$ as $m \to 0$.

To this end, observe that

$$\begin{aligned}
&\left(\frac{1}{[\varphi(\mu(B_2))]^p \mu(B)} \int_B |f(y) - f_B|^p \, d\mu(y)\right)^{\frac{1}{p}} \\
&= \left(\frac{1}{[\varphi(\mu(B_2))]^p \mu(B)} \int_B \left|\sum_{k=1}^{\infty} f_k(y) - (f_k)_B\right|^p d\mu(y)\right)^{\frac{1}{p}} \\
&\le \sum_{k=1}^{\infty} \left(\frac{1}{[\varphi(\mu(B_2))]^p \mu(B)} \int_B |f_k(y) - (f_k)_B|^p \, d\mu(y)\right)^{\frac{1}{p}} \\
&\le \sum_{k=1}^{\infty} \|f_k\|_{\mathrm{BMO}_\varphi^{(p)}} < \infty,
\end{aligned}$$

thus $\|f\|_{\mathrm{BMO}_\varphi^{(p)}} < \infty$, and then $f \in \mathrm{BMO}_\varphi^{(p)}(X)$. This proves part (a).

On the other hand,

$$\begin{aligned}
&\left(\frac{1}{[\varphi(\mu(B_2))]^p \mu(B)} \int_B \left|\left(\sum_{k=1}^{\infty} f_k - f\right)(y) - \left(\sum_{k=1}^{m} f_k - f\right)_B\right|^p d\mu(y)\right)^{\frac{1}{p}} \\
&= \left(\frac{1}{[\varphi(\mu(B_2))]^p \mu(B)} \int_B \left|\sum_{k=m+1}^{\infty} (f_k(y) - (f_k)_B)\right|^p d\mu(y)\right)^{\frac{1}{p}} \\
&\le \sum_{k=m+1}^{\infty} \left(\frac{1}{[\varphi(\mu(B_2))]^p \mu(B)} \int_B |f_k(y) - (f_k)_B|^p \, d\mu(y)\right)^{\frac{1}{p}} \\
&\le \sum_{k=m+1}^{\infty} \|f_k\|_{\mathrm{BMO}_\varphi^{(p)}} \to 0, \quad \text{as } m \to \infty.
\end{aligned}$$

Hence $\| \sum_{k=1}^{\infty} f_k - f \|_{\mathrm{BMO}_\varphi^{(p)}} \to 0$ as $m \to 0$. This proves part (b), completing the proof of the theorem. □

Let us finish this chapter with the following remarkable result.

Theorem 6.3. *Let $f \in \mathrm{BMO}_\varphi^{(p)}$, then there is a constant $\epsilon > 0$ such that*

$$\sup \mu(\{x \in B : |f(x) - f_B| > \lambda\})/\mu(B) \le e^{-\lambda/\epsilon}, \tag{6.9}$$

where $\lambda > \lambda(\epsilon, f)$. Indeed, by Theorem 6.1, we have $\epsilon = C\|f_k\|_{\mathrm{BMO}_\varphi^{(p)}}$ and $\lambda(\epsilon, f) = C\|f_k\|\mathrm{BMO}_\varphi^{(p)}$. Moreover, if

$$\epsilon(f) = \inf\{\epsilon : (6.9)\ holds\},$$

then

$$\frac{\epsilon(f)}{e\varphi(\mu(B))} = \lim_{p\to\infty} \frac{1}{p} \|f\|_{\mathrm{BMO}_\varphi^{(p)}}.$$

Proof. Since

$$\begin{aligned}
\int_{B(x,r)} |f(x) - f_B|^p \, d\mu(x) &= p \int_0^\infty \lambda^{p-1} \mu(\{x \in B : |f(x) - f_B| > \lambda\}) \, d\lambda \\
&\le p\mu(B) \int_0^\infty \lambda^{p-1} e^{-\lambda/\epsilon} \, d\lambda \\
&= \mu(B)\epsilon^p \int_0^\infty u^{p-1} e^u \, du.
\end{aligned}$$

Thus,

$$\frac{1}{\mu(B)} \int_{B(x,r)} |f(x) - f_B|^p \, d\mu(x) \le \epsilon^p p\Gamma(p).$$

Next, we obtain

$$\frac{1}{p} \sup\left(\frac{1}{[\varphi(\mu(B))]^p \mu(B)} \int_B |f(y) - f_B|^p \, d\mu(y) \right)^{\frac{1}{p}} \le \frac{\epsilon[p\Gamma(p)]^{\frac{1}{p}}}{\varphi(\mu(B))p}$$

and then

$$\lim_{p\to\infty} \frac{1}{p} \sup\left(\frac{1}{[\varphi(\mu(B))]^p \mu(B)} \int_B |f(y) - f_B|^p \, d\mu(y) \right)^{\frac{1}{p}} \le \frac{\epsilon(f)}{e\varphi(\mu(B))}. \tag{6.10}$$

On the other hand, if $\epsilon < \epsilon(f)$, then there exists $B_0 \subset X$ such that

$$e^{-\lambda/\epsilon} \le \mu(\{x \in B_0 : |f(x) - f_B| > \lambda\})/\mu(B_0).$$

Thus

$$p\mu(B_0)\int_0^\infty \lambda^{p-1}e^{\lambda/\epsilon}\,d\lambda < p\int_0^\infty \lambda^{p-1}\mu(\{x \in B : |f(x) - f_B| > \lambda\})\,d\lambda$$

and

$$\frac{\epsilon[\Gamma(p)]^{\frac{1}{p}}}{\varphi(\mu(B))p} < \frac{1}{p}\left(\frac{1}{[\varphi(\mu(B))]^p\mu(B)}\int_B |f(y) - f_B|^p\,d\mu(y)\right)^{\frac{1}{p}}.$$

It is follows that

$$\frac{\epsilon(f)}{e\varphi(\mu(B))} < \lim_{p\to\infty}\frac{1}{p}\sup\left(\frac{1}{[\varphi(\mu(B))]^p\mu(B)}\int_B |f(y) - f_B|^p\,d\mu(y)\right)^{\frac{1}{p}}. \tag{6.11}$$

Combining (6.10) and (6.11), we obtain the desired result. □

Remark 6.2. Theorem 6.3, together with Lemma 6.2, allows us to estimate the distance from $\mathrm{BMO}_\varphi^{(p)}$ to L_∞. In other words, we can estimate

$$\inf_{g\in L_\infty} \|f - g\|_{\mathrm{BMO}_\varphi^{(p)}}$$

with $f \in \mathrm{BMO}_\varphi^{(p)}$.

Index

https://doi.org/10.1515/9783112223246-007

Bibliography

[1] M. A. Ariño and B. Muckenhoupt. Maximal functions on classical Lorentz spaces and Hardy's inequality with weights for nonincreasing functions. *Transactions of the American Mathematical Society*, 320(2):727–735, 1990.

[2] C. Bennet and R. Sharpley. *Interpolation of operator*. Academic Press, 1998.

[3] R. P. Boas. Some uniformly convex spaces. *Bulletin of the American Mathematical Society*, 46(4):304–311, 1940.

[4] A. P. Calderón and A. Zygmund. On the existence of certain singular integrals. *Acta Mathematica*, 88:85–139, 1952.

[5] T. Carleman. Sur les fonctions quasi-analytiques (comptes rendus du ve congres des mathematiciens scandinaves, 1922). *Helsingfors*, pages 181–196, 1922.

[6] F. Carlson. Une inégalité. *Arkiv för Matematik, Astronomi och Fysik*, 25B(1):1–5, 1934.

[7] M. Carro, L. Pick, J. Soria, and V. D. Stepanov. On embeddings between classical Lorentz spaces. *Mathematical Inecualities and Applications*, 4:397–428, 2001.

[8] M. Carro and J. Soria. Weighted Lorentz spaces and the Hardy operator. *Journal of Functional Analysis*, 112(2):480–494, 1993.

[9] R. E. Castillo. Nonlinear bessel potentials and generalizations of the Kato class. *Proyecciones*, 30(3):285–294, 2011.

[10] R. E. Castillo. *Espacios L_p*. Universidad Nacional de Colombia, Bogotá D. C., 2014.

[11] R. E. Castillo and H. C. Chaparro. Weighted composition operator on two dimensional Lorentz sapces. *Mathematical Inequalities & Applications*, 20(3):773–799, 2017.

[12] R. E. Castillo, H. C. Chaparro, and J. C. Ramos-Fernández. Orlicz-Lorentz spaces and their composition operators. *Proyecciones (Antofagasta)*, 34(1):85–105, 2015.

[13] R. E. Castillo, H. C. Chaparro, and J. C. Ramos-Fernández. Orlicz-Lorentz spaces and their multiplication operators. *Hacettepe Journal of Mathematics and Statistics*, 44(5):991–1009, 2015.

[14] R. E. Castillo, R. Humberto, J. C. Ramos-Fernández, and M. Salas-Brown. Multiplication operator on Köthe spaces: measure of non-compactness and closed range. *Bulletin of the Malaysian Mathematical Sciences Society*, 42(4):1523–1534, 2019.

[15] R. E. Castillo, R. León, and E. Trousselot. Multiplication operator on $L_{(p,q)}$ spaces. *Panamerican Mathematical Journal*, 19(1):37–44, 2009.

[16] R. E. Castillo and H. Rafeiro. *An introductory course in Lebesgue spaces*. Springer, New York, 2016.

[17] R. E. Castillo, H. Rafeiro, and E. M. Rojas. Unique continuation of the quasilinear elliptic equation on Lebesgue spaces l_p. *Azerbaijan Journal of Mathematics*, 11(1):136–156, 2021.

[18] R. E. Castillo, J. C. Ramos-Fernández, and E. Trousselot. Functions of bounded (φ, p) mean oscillation. *Proyecciones*, 27(2):185–200, 2008.

[19] R. E. Castillo, R. Sánchez, and E. Trousselot. Weighted composition operator on the gamma spaces $\Gamma_X(w)$. *Analysis and Mathematical Physics*, 11(4):161, 2021.

[20] R. E. Castillo, C. Suarez, and E. Trousselot. Composition operator on $\Lambda_X^p(x)$ spaces. *Quaestiones Mathematicae*, 40(6):833–848, 2017.

[21] R. E. Castillo and E. Trousselot. A simple proof of a weighted inequality for the Hardy-Littlewood maximal operator in $\mathbb{R}^n$. *Acta Mathematica Academiae Paedagogicae Nyíregyháziensis (New Series)*, 25(2):241–245, 2009.

[22] R. E. Castillo, F. A. Vallejo, and J. C. Ramos-Fernández. Multiplication and composition operators on weak L_p spaces. *Bulletin of the Malaysian Mathematical Sciences Society*, 38(3):927–973, 2015.

[23] H. C. Chaparro. *Weighted composition operator on multidimensional Lorentz spaces and glimpse on multipliers between bounded p-variation spaces*. PhD thesis, Universidad Nacional de Colombia, 2018.

[24] O. A. Chaparro. *Weighted multiplication operator on weighted Lebesgue sequence spaces*. Master's thesis, Universidad Nacional de Colombia, 2025. Available at https://repositorio.unal.edu.co/handle/unal/88121.

https://doi.org/10.1515/9783112223246-008

[25] F. Chiarenza and M. Frasca. A remark on a paper by C. Fefferman. *Proceedings of the American Mathematical Society*, 108:407–409, 1990.

[26] J. A. Clarkson. Uniformly convex spaces. *Transactions of the American Mathematical Society*, 40(3):396–414, 1936.

[27] R. R. Coifman and C. Fefferman. Weighted norm inequalities for maximal functions and singular integrals. *Studia Mathematica*, 51:241–250, 1974.

[28] R. R. Coifman and Y. Meyer. *Au delà des opérateurs pseudo-différentiels*, volume 57 of *Astérisque*. Société Mathématique de France (SMF), Paris, 1978.

[29] D. Cruz-Uribe and A. Fiorenza. *Variable Lebesgue spaces. Foundations and harmonic analysis. Applied and Numerical Harmonic Analysis*. Birkhäuser/Springer, New York, NY, 2013.

[30] D. Cruz-Uribe, O. M. Guzmán, and H. Rafeiro. Weighted Riesz bounded variation spaces and the Nemitskii operator. *Azerbaijan Journal of Mathematics*, 10(2):125–139, 2020.

[31] D. Cruz-Uribe, J. M. Martell, and C. Pérez. *Weights Extrapolation and the theory of Rubio de Francia*, volume 215 of *Operator Theory: Advance and Applications*. Birkhäuser/Springer Basel AG, Basel, 2011.

[32] D. Danielli, N. Garofalo, and D. Nhieu. Trace inequality for Carnot-Carathéodory space and applications. *Annali della Scuola Normale Superiore di Pisa*, 27:195–252, 1998.

[33] I. Daubechies. *Ten lectures on wavelets*, volume 61 of *CBMS-NSF Regional Conference Series in Applied Mathematics*. SIAM, Philadelphia, PA, 1992.

[34] G. Di Fazio and P. Zamboni. A Fefferman-Poincaré type inequality for Carnot-Carathéodory vector fields. *Proceedings of the American Mathematical Society*, 130(9):2655–2660, 2002.

[35] D. Edmunds and W. Evans. *Hardy-type operators*, pages 11–61. Springer Berlin Heidelberg, Berlin, Heidelberg, 2004.

[36] C. Fefferman. The uncertainty principle. *Bulletin of the American Mathematical Society*, 9(2):129–206, 1983.

[37] J. García Cuerva and J. L. Rubio de Francia. *Weigthed norm inequalities and related topics*, volume 46. North-Holland, 1985.

[38] J. B. Garnett and P. W. Jones. The distance in BMO to L^∞. *Annals of Mathematics. Second Series*, 108:373–393, 1978.

[39] F. W. Gehring. The L^p integrability of partial derivatives of a quasiconformal mapping. *Acta Mathematica*, 130:265–277, 1973.

[40] L. Grafakos. *Classical Fourier analysis*, volume 249 of *Graduate Texts in Mathematics*. Springer, New York, second edition, 2008.

[41] G. H. Hardy. Note on a theorem of Hilbert. *Math. Zeit.*, 6(3):314–317, 1920.

[42] G. H. Hardy. Note on a theorem of Hilbert concerning series of positive terms. *Proceedings of the London Mathematical Society*, 23(2):45–46, 1925.

[43] G. H. Hardy. A maximal theorem with function theoretic applications. *Acta Mathematica*, 54:81–116, 1930.

[44] G. H. Hardy. A note on two inequalities. *Journal of the London Mathematical Society*, s1-11(3):167–170, 1936.

[45] G. H. Hardy and J. E. Littlewood. Some properties of fractional integrals. *Mathematische Zeitschrift*, 27:565–606, 1927.

[46] D. Hilbert. Grundzüge einer allgemeinen theorie der linearen integralgleichungen. *Göttinger Nachrichten*, pages 157–227, 1906.

[47] R. A. Hunt. On $L(p, q)$ spaces. *L'Enseignement Mathématique*, 12(2):249–276, 1996.

[48] R. A. Hunt and B. Muckenhoupt. Weighted norm inequalities for the conjugate function and Hilbert transform. *Transactions of the American Mathematical Society*, 176:227–251, 1973.

[49] F. John and L. Nirenberg. On functions of bounded mean oscillation. *Communications on Pure and Applied Mathematics*, 14:415–426, 1961.

[50] E. Kreyszig. *Introductory functional analysis with applications*. John Wiley & Sons, Windsor, 1978.

[51] A. Kufner, L. Maligranda, and L. Persson. The prehistory of the Hardy inequality. *American Mathematical Monthly*, 113(8):715–732, 2006.
[52] L. Larsson. *Carlson type inequalities and their applications*. PhD thesis, Matematiska institutionen, 2003.
[53] L. Larsson, L. Maligranda, L. Persson, and J. Pecaric. *Multiplicative inequalities of Carlson type and interpolation*. World Scientific, New Jersey, 2006.
[54] G. G. Lorentz. Some new functions spaces. *Annals of Mathematics*, 51:37–55, 1950.
[55] G. G. Lorentz. On the theory of spaces Λ. *Pacific Journal of Mathematics*, 1:411–429, 1951.
[56] R. A. Macías and C. Segovia. Lipschitz functions on spaces of homogeneous type. *Advances in Mathematics*, 33:257–270, 1979.
[57] J. Mateu, P. Mattila, A. Nicolau, and J. Orobitg. BMO for nondoubling measures. *Duke Mathematical Journal*, 102(3):533–565, 2000.
[58] B. Muckenhoupt. Weighted norm inequalities for the Hardy maximal function. *Transactions of the American Mathematical Society*, 165:207–226, 1972.
[59] B. Muckenhoupt and R. L. Wheeden. Weighted norm inequalities for fractional integrals. *Transactions of the American Mathematical Society*, 192:261–274, 1974.
[60] O. Ogur. Some geometric properties of weighted Lebesgue spaces $L_p(g)$. *Facta Universitatis. Series: Mathematics and Informatics*, 33(4):523–530, 2018.
[61] M. Rosenblum. Summability of Fourier series in $L^p(d\mu)$. *Transactions of the American Mathematical Society*, 105:32–42, 1962.
[62] J. L. Rubio de Francia. Weighted norm inequalities and vector valued inequalities. *Harmnic analysis, Proc. Conf., Minneapolis 1981*, volume 908 of *Lecture Notes in Mathematics*, pages 86–101, 1982.
[63] M. Schechter. *Spectra of partial differential operators*. North-Holland Publishing Co., Amsterdam, New York, 1986.
[64] J. M. Steele. *The Cauchy-Schwarz master class: an introduction to the art of mathematical inequalities*. Cambridge University Press, New York, 2004.
[65] E. M. Stein. *Singular integrals and differentiability properties of functions (PMS-30)*. Princeton University Press, 1971.
[66] A. Torchinsky. *A real variable methods in harmonic analysis*, volume 179. Academic press, New York, 1985.
[67] P. Zamboni. Unique continuation for non-negative solutions of quasilinear elliptic equations. *Bulletin of the Australian Mathematical Society*, 64:149–156, 2001.

www.ingramcontent.com/pod-product-compliance
Lightning Source LLC
Chambersburg PA
CBHW081538220326
41598CB00036B/6479
* 9 7 8 3 1 1 9 1 4 2 2 0 5 *